T0180788

Green Energy and Technology

For further volumes:
http://www.springer.com/series/8059

Vesna Žegarac Leskovar
Miroslav Premrov

Energy-Efficient
Timber-Glass Houses

 Springer

Vesna Žegarac Leskovar
Miroslav Premrov
University of Maribor
Maribor
Slovenia

ISSN 1865-3529 ISSN 1865-3537 (electronic)
ISBN 978-1-4471-7211-6 ISBN 978-1-4471-5511-9 (eBook)
DOI 10.1007/978-1-4471-5511-9
Springer London Heidelberg New York Dordrecht

Translation: Danijela Žegarac
Picture design: Anja Patekar

Printed on acid-free paper

Springer is part of Springer Science+Business Media (www.springer.com)

Contents

Chapter 1
Introduction

Abstract The introductory chapter sets a background frame and reveals the main reasons which encouraged the authors into researching the topic of energy efficiency of buildings. Section 1.2 is a brief overview of the authors' activities in the fields of energy efficiency and timber-glass construction, while Sect. 1.3 shortly outlines the content of the book.

1.1 Why Dealing with the Topic of Timber-Glass Buildings?

Climate changes of the last few decades do not only encourage researches into the origins of their onset, but they also mean a warning and an urgent call for a need to remove their causes and alleviate the consequences affecting the environment. Construction is, besides the fields of transport and industry, one of the main users of the prime energy from fossil sources, which makes this sector highly responsible for the implementation of climate-environmental policies. Activities linked to energy efficiency and the related use of renewable sources of energy are not infrequent in Slovenia, nevertheless, the fields of architecture and construction still offer numerous possibilities of reaching the goals set by directives on energy efficiency in buildings. Looking for alternative, eco-friendly solutions in residential and public building construction remains our most vital task, whose holistic problem solving requires knowledge integration. The present book represents merely a piece in the jigsaw of different kinds of knowledge that will need to undergo mutual integration and upgrading in order to be used in designing an optimal energy-efficient timber-glass building.

The current work can be useful to designers and future experts in their planning of optimal energy-efficient timber-glass buildings. The study is based on using timber and glass which used to be rather neglected as construction materials in certain historical periods. Nevertheless, timber achieved recognition as one of the

V. Žegarac Leskovar and M. Premrov, *Energy-Efficient Timber-Glass Houses*,
Green Energy and Technology, DOI: 10.1007/978-1-4471-5511-9_1,
© Springer-Verlag London 2013

oldest building materials in different countries worldwide. With the appearance of cast and wrought iron in the eighteenth century along with the subsequent use of reinforced concrete and steel in the twentieth century, which all enabled mass production and construction of larger structural spans, timber lost its dominance as a building material McLeod [1]. Only in recent decades has timber been rediscovered, partly due to the contemporary manufacture of prefabricated timber elements and partly owing to high environmental potential of this renewable natural building material.

Although glass has been used to enclose space for nearly two millennia, the roots of modern glass construction reach back to the nineteenth century green houses in England, witnessing one of the first instances of using glass as a load-bearing structural element in combination with the iron skeleton, Wurm [2]. Throughout the twentieth century, glass was no longer used as load-bearing element, but rather as an aesthetic element of the building skin with strongly emphasized potential of transparency enabling natural lighting and visual contact of the interior and exterior space. In contrast to the listed positive properties, glass used to be treated as the weakest point of the building envelope from the thermal point of view. Dynamic evolution of the glazing in the last 40 years resulted in insulating glass products with highly improved physical and strength properties, suitable for application in contemporary energy-efficient buildings, not only as material responsible for solar gains and daylighting, but also as a component of structural resisting elements.

With suitable technological development and appropriate use, timber and glass are nowadays becoming essential construction materials as far as the energy efficiency is concerned. Their combined use is extremely complicated, from both the constructional point of view as well as from that of energy efficiency and sets multiple traps for designers. Moreover, a novelty value of modern glass is seen in its being treated as a load-bearing material replacing the elements (diagonal elements, sheathing boards) which normally provide horizontal stability of timber structures. A good knowledge of advantages and drawbacks of timber-glass structures is thus vitally important.

1.2 Authors' Work in the Field of Energy Efficiency and Timber-Glass Construction

Within a selection of most important issues, our activities in the frames of the University of Maribor, Faculty of Civil Engineering, focus primarily on research work and its application into practice, on educating students and the broader public (Fig. 1.1). Our scientific work in the field of energy efficiency of the buildings concentrates on researching design models of energy-efficient timber-glass buildings, which combines the knowledge of architecture, timber-glass construction and building physics. We strive to link the findings of our research work with practice

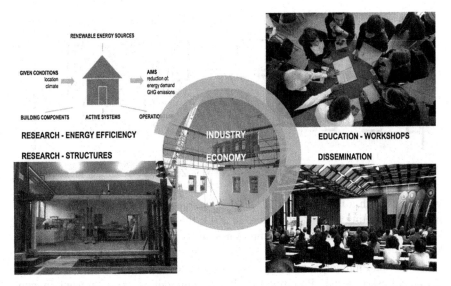

Fig. 1.1 Scheme of different activities carried out by the University of Maribor, Faculty of Civil Engineering, linked to demands arising from the construction industry and economy

via cooperating with the relevant branches of business, with Slovene prefabricated timber-frame house manufacturers who realize the vitality of making progress in the field of timber construction. A considerable part of the civil engineering business is said to have become environmentally aware and, following the demands of the modern market, also well informed as far as the basics of energy-efficient construction are concerned, which leads us to consider the importance of awareness building among the broader public sector, end users of energy-efficient buildings and particularly among future experts—current students who will use their knowledge in practice and see to its upgrading and further expansion. Consequently, we transfer scientific research findings into the process of education as well as into organization of expert meetings discussing energy efficiency in the domains of civil engineering and architecture. Through such education and by informing experts and future designers, we participate in broadening the knowledge as well as in building awareness of the importance of eco-friendly design approaches.

1.2.1 Students' Workshops on Timber-Glass Buildings

Creating design ideas for timber-glass energy-efficient buildings was the central point of study workshops carried out from 2010 to 2012. Starting with projects for a single-family timber house in 2010, the first step was to inform students about the basics of energy efficiency, Žegarac Leskovar et al. [3]. The follow-up

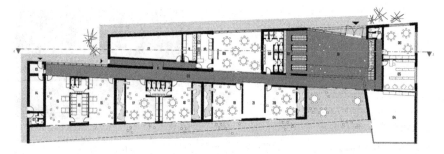

Fig. 1.2 Ground floor plan of the Sovica Kindergarten, Žegarac Leskovar and Premrov [4]

workshops with a focus on public building design were marked with more complexity, presenting a logical upgrade. In 2011, the participating students designed kindergartens and multi-purpose buildings for a small community of Destrnik, Slovenia, Žegarac Leskovar and Premrov [4]; while in 2012, they dealt with residential and municipal buildings for another Slovenian community, Podlehnik, Žegarac Leskovar and Premrov [5]. Both communities are interested in constructing one of the buildings designed by our students.

1.2.1.1 The Sovica Kindergarten, a Project for the Community of Destrnik

The building is divided into two parts which slide past each other. The result is a compact form, which functionally divides the kindergarten into classrooms and other areas (Figs. 1.2 and 1.3).

The kindergarten is designed in the timber-frame panel structural system. The average U-value of the thermal envelope is 0.10 W/m^{2}K. Although glass is not treated as a load-bearing material, it has to be designed with utmost care in order to benefit from the solar gain potential. The selected configuration of the façade

Fig. 1.3 Model of the Sovica Kindergarten, Žegarac Leskovar and Premrov [4]

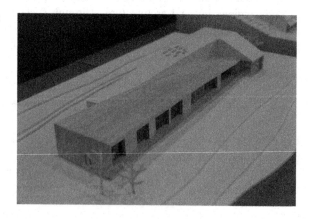

glazing is 4E-12-4-12-E4 with $U_g = 0.51$ W/m^{2}K, $g = 52$ % and $U_f = 0.73$ W/m^{2}K. The share of glazing in the south-oriented façade is 45 %. Roof glazing is additionally integrated in order to transfer natural light into spaces which are not sufficiently daylight through the façade glazing. The shape factor of the building is 0.7 m^{-1}, indicating a relatively compact form. The building is equipped with active technical systems, a heat pump and a cooling unit. The calculated annual heating demand is 14 kWh/m^{2}a.

Authorship of the University of Maribor, Faculty of Civil Engineering:
Architectural design: Maša Kresnik, Sanda Moharič, Tajda Potrč and Anja Patekar, Students of Architecture;
Structural design: Anja Pintarič and Maruša Retužnik, Students of Civil Engineering;
Supervision: Assistant Professor Vesna Žegarac Leskovar, Architect and Professor Miroslav Premrov, Civil Engineer.

1.2.1.2 Single-Family Passive House Marles

In 2012, a local national company Marles Hiše Maribor, a manufacturer of timber-frame panel houses, commissioned a study for the development of different innovative models of timber-glass energy-efficient houses among which they selected the best prototype with the intention of production and market launch. The house presented in Fig. 1.4 is a winning project, which is currently being offered as a standard prefabricated house type of the Marles company, Marles [6].

The single-family house with a ground floor area of 150 m^2 is constructed in the timber-frame panel system with the average U-value of opaque elements of the thermal envelope being 0.1 Wm^{2}K. The glazing surfaces in the façade and roof allow for an optimal daylight and comfortable indoor living climate. The active technical systems are a compact unit with a heat pump and heat recovery ventilation. The calculated energy demand for space heating is 15 kWh/m^{2}a.

Fig. 1.4 Visualization of the single-family house Marles

Authorship of the University of Maribor, Faculty of Civil Engineering:
Architectural design: Maša Kresnik and Sanda Moharič, Students of Architecture;
Structural design: Miha Pukšič, Students of Civil Engineering;
Energy design: Klara Mihalič, Anže Rosec, students of Civil Engineering;
Supervision: Assistant Professor Vesna Žegarac Leskovar, Architect, Professor
Miroslav Premrov and Assistant Professor Erika Kozem Šilih, Civil Engineers.

1.3 The Content of the Book

The current book has been written in order to present the importance of combining
two basic design approaches, the architectural and structural, both focusing on
energy-efficient problem solving.

The book consists of four chapters. This chapter is an introduction to the topic
of energy-efficient timber-glass buildings. It explains the importance of the relevant integration of the sciences of architecture and civil engineering on the one
hand and that of research and academic work along with the newest trends and
requirements of modern timber-glass construction, on the other. Chapter 2 presents
the basic principles of energy-efficient design, with the focus on parameters that
influence the energy performance of a building. Timber's material characteristics
are accurately described in Chap. 3 which also lists general types of timber
structural systems, describes computational models and methods and discusses
stability problems. Chapter 4 presents material characteristics of glass along with
the research results related to the combined use of timber and glass from the
viewpoint of energy and structural stability.

We hope the findings of this book will act as beneficial encouragement and
inspiration for further researches and for more successful energy-related problem
solving in Europe and elsewhere.

References

1. McLeod V (2009) Detail in contemporary timber architecture. Laurence King Publishing Ltd
2. Wurm J (2007) Glass structures: design and construction of self-supporting skins. Birkhäuser
   Verlag AG Basel-Boston_Berlin
3. Žegarac Leskovar V, Premrov M, Lukič M, Vene Ž (2010) Delavnica Lesena nizkoenergijska
   hiša (Workshop on Timber Low-energy House). E-zavod, Zavod za celovite rešitve
4. Žegarac Leskovar V, Premrov M (2011) Architectural design approach for energy-efficient
   timber frame public buildings. University of Maribor, Maribor Faculty of Civil Engineering
5. Žegarac Leskovar V, Premrov M (2013) Educational projects on energy-efficient timber
   buildings, architectural workshop for the municipality of Podlehnik. University of Maribor,
   Slovenia (Faculty of Civil Engineering)
6. Marles (2013) Marles future: promotional catalogue of the Marles company

Chapter 2
Energy-Efficient Building Design

Abstract The current chapter discusses a number of important aspects whose influence on the energy efficiency of new buildings calls for their careful consideration as early as in the design phase. With the basics of energy-efficient building design figuring in Sect. 2.1, the next important topic contained in Sect. 2.2 deals with commonly used classification systems determining the energy efficiency level of buildings. In order to understand energy-efficient design principles, basic facts on energy flows in buildings are given in Sect. 2.3. The relation between the building design, climatic influences and the building site analysis can be found in Sect. 2.4. Section 2.5 introduces a set of main design parameters, such as orientation, shape of the building, zoning of interior spaces and the building components. Description of the building components focuses mainly on those composing the building thermal envelope, with glazing surfaces and timber construction being only briefly presented, while a more detailed specification of the two materials follows in Chaps. 3 and 4. For the complexity of energy-efficient design, passive design strategies comprising passive solar heating, cooling, ventilation and daylighting are considered in Sect. 2.6. Finally, Sect. 2.7 provides an overview of the role of active technical systems, since they have become an indispensable constituent element of contemporary energy-efficient houses.

2.1 Basics of Energy-Efficient Building Design

Climate-conscious architecture, bioclimatic architecture or energy-efficient architecture are commonly used terms presenting specific approaches in contemporary architectural building design, which have to be applied in conjunction with the structural, technical and aesthetic aspects of architecture. Nevertheless, general guidelines related to building design along with its relation to the environment are not a novelty since they are to some extent based on vernacular architectural principles having existed for centuries. On the other hand, the reflection of the current global energy situation is seen in the demand for energy-efficient building

V. Žegarac Leskovar and M. Premrov, *Energy-Efficient Timber-Glass Houses*,
Green Energy and Technology, DOI: 10.1007/978-1-4471-5511-9_2,
© Springer-Verlag London 2013

design to be determined by precisely defined parameters which affect the energy balance of buildings (Fig. 2.1).

Energy-efficient building design requires a careful balance of the energy consumption, energy gain and energy storage. As shown in Fig. 2.1, the basic design principle integrates the building components into a system taking maximum advantage of the building's environment, climatic conditions and available renewable energy sources. The aim is to reduce the need for conventional heating and ventilation systems, which are inefficient and consume fossil energy sources. The use of contemporary active technical systems exploiting renewable energy is therefore advised instead. Apart from higher energy efficiency and reduced environmental burdening, energy-efficient building design results in a comfortable indoor climate, which is of utmost importance for the occupants' well-being. The occupants play an important role in the system of energy-efficient buildings, since only with proper use can the buildings' energy balance reach a level planned by the engineers.

Planning and designing energy-efficient buildings is a complex process whose definition could be understood as a three-levelled one. The first basic design level comprises an optimum selection of the building components, i.e., the structural design concept, thermal envelope composition, construction details, type of glazing and other materials, with respect to the location, climatic data and a suitable orientation. The following level is that of passive design strategies which allow for heating with solar gains, cooling with natural ventilation, using thermal mass for energy storage where renewable energy sources are exploited with no need for electricity. Only the third, i.e., the last level involves design concepts of the building's active technical systems using renewable sources of energy with the necessary recourse to electrical energy. Efficient planning and design of buildings aims at skilfulness and originality of design concepts at the first two levels to the extent where the need for active systems arises within the least possible degree.

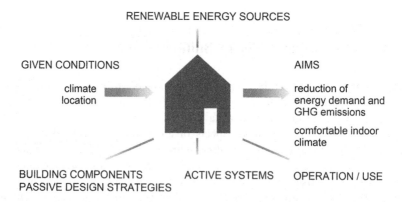

Fig. 2.1 Basic principle of energy-efficient design

2.2 Classification of Buildings According to Energy Efficiency

Energy efficiency requirements in building codes can ensure that concern for energy efficiency is taken in the design phase, which leads to realization of large potentials for achieving good energy efficiency standards in new buildings [19]. Currently, there exists no common classification of energy-efficient buildings. On the contrary, there is a variety of standards and energy labels used in the construction industry across Europe with commonly appearing terms, such as low-energy house, passive house, zero carbon house, zero energy house, 3-litre house etc., [9].

To determine specific energy standards for buildings, many European countries use classification systems based on national building codes or recur to launching special labels, e.g., the Swiss Minergie, German Passive House or Austrian Klima:aktiv House. In order to achieve a certain label or be classified into a specific standard determined by the building codes, a building has to satisfy specific-energy consumption-related requirements which are not uniform to all classification systems. The existing classification systems vary mainly in the type of energy taken into account. Certain systems focus merely on the specific energy demand for space heating while others also consider CO_2 emissions or even respect additional types of energy use, for instance primary energy demand. Only rarely will classification systems take into consideration a complex set of energy indicators integrating all types of the energy use in buildings, where apart from the ones mentioned above, the energy use for space cooling, water heating, air conditioning, consumption of electricity, etc., proves to be of vital importance. Furthermore, some classification systems introduce separate measures for residential and public buildings, for small and multi-storey buildings, for new and renovated buildings. A number of countries established requirements modified for different regions. Generally, a certain energy standard is achieved through structural measures and active technical systems. User behaviour has no effect on the standard, although it does affect the actual energy consumption.

To illustrate the requirements for different standards, a few of the existing classification systems are presented in Tables 2.1, 2.2 and 2.3.

The data in Table 2.1 show variations of two national systems used to determine the building's energy label. The goal referring to the maximum energy demand for heating in new buildings set in the Slovene National Building Regulations is around 50–60 kWh/m^{2}a with that of the Austrian National Building Regulations being below 50 kWh/m^{2}a. A building satisfying the above goals can be treated as energy-efficient; however, the aim is to design buildings in a manner to use less energy than defined by the maximum value set in the building codes. Instead of classes A, B and C, a descriptive terminology based on labels is more widely used in the construction industry. The rates of energy consumption in buildings commonly used in Austria are presented in Table 2.2.

Table 2.1 Slovene classification of the building's energy efficiency (Rules on the methodology of construction and issuance of building energy certificates 2010) and Austrian classification of the building's energy efficiency [20, 23]

Energy class	Annual heating demand Q_h [kWh/m²a]	Energy class	Annual heating demand Q_h [kWh/m²a]
A1	0–10	A++	≤10
A2	10–15	A+	10–15
B1	15–25	A	15–25
B2	25–35	B	25–50
C	35–60	C	50–100
D	60–105	D	100–150
E	105–150	E	150–200
F	150–210	F	200–250
G	210–300	G	250–300

Table 2.2 Austrian classification of the building's energy efficiency on the basis of commonly used labels for construction industry

Label	Annual heating demand Q_h [kWh/m²a]	
	Small buildings <130 m²	Multi-storey buildings
Passive house	≤10	≤10
Super low-energy house	10–36	10–20
Low-energy house	36–45	20–25

The above classification has separate requirements for multi-storey buildings and smaller buildings whose floor area is lower than 130 m².

The German classification system, based on the requirements defined by Energieeinsparverordnung für Gebäude 2009, EnEV [8], is completely different. Buildings are rated according to the level of improvement of energy performance determined in EnEV [8]. It should be noted that the amended Energieeinsparverordnung für Gebäude the EnEV 2014 is currently under adoption. On the other hand, the Passive House Institute registered their own label for buildings with special design requirements called the "Passive House Certificate" [10]. Table 2.3 shows a selection of energy standards currently applied in Germany.

In general, the existing classification systems are based on the directives which are modified within a specified time period. Some of the systems presented in Tables 2.1, 2.2 and 2.3 arise from the building codes based on the European directive on energy performance of buildings which requires classification of all new buildings according to the energy certificate whose validity lasts for 10 years.

The current section deals mainly with factors which produce influence on the energy efficiency of buildings and can be taken into consideration in the design stage. The level of energy efficiency can be calculated by using the existing software tools. In order to understand energy-efficient design principles presented later in this chapter, some basic knowledge on energy flows in buildings is required.

Table 2.3 German classification of the building's energy efficiency

Energy class	Annual heating demand Q_H [kWh/m^{2}a]	Primary energy demand Q_P [kWh/m^{2}a]	Final energy demand Q_e [kWh/m^{2}a]	Heat transmission losses H'_T
Plus energy house		≤ 0	≤ 0	
Passive house	≤ 15	$\leq 120^{**}$		
*KfW—energy-efficient house 40 [8]		≤ 40 % of the ***maximum value set in EnEV 2009		≤ 55 % of the maximum value set in EnEV 2009
KfW—energy-efficient house 55 [8]	≤ 35	≤ 55 % of the maximum value set in EnEV 2009		≤ 70 % of the maximum value set in EnEV 2009
KfW—energy-efficient house 70 [8]	≤ 45	≤ 70 % of the maximum value set in EnEV 2009		≤ 85 % of the maximum value set in EnEV 2009
KfW—energy-efficient house 85 [8]	≤ 55	≤ 85 % of the maximum value set in EnEV 2009		≤ 100 % of the maximum value set in EnEV 2009

German classification of the building's energy efficiency on the basis of the Energieeinsparverordnung für Gebäude 2009, EnEV [8], Förderstufen der KfW Bankengruppe and Passive house [10], definitions
* KfW—Funding levels of the KfW Bank Group
** Maximum value for Q_p set in Feist [10] contains the requirements for heating, water heating, cooling, ventilation and household electricity
*** Maximum value for Q_p set in EnEV is approximately 60 kWh/m^2 a and contains no demand for household electricity

2.3 Energy Flows in Buildings

A building can be considered as a thermal system with a series of heat flows, inputs and outputs [24], such as the transmission heat losses or gains (Q_t), ventilation heat losses or gains (Q_v), internal heat gains (Q_i) and solar heat gains (Q_s). The thermal response of the building is preconditioned by the relationship between the heat gains and losses, where the sum of all energy flows results in the amount of energy (ΔQ) that has to be supplied to or extracted from the building in order to reach a comfortable indoor living climate. If the sum of all heat flows is zero, the building is reaching the thermal balance.

The following equation shows the main heat flows in a building exerting influence on the indoor living climate:

$$Q_t + Q_v + Q_i + Q_s = \Delta Q \qquad (2.1)$$

where the main quantities are the following:

Q_t transmission heat losses or gains caused by heat flow through the elements of the building envelope,

Q_v ventilation heat losses or gains caused by air exchange between the building and its surrounding (air infiltration, natural ventilation, mechanical ventilation, air leakage through the building envelope),

Q_i internal heat gains generated inside the building by occupants, lighting and household appliances,

Q_s solar heat gains caused by solar radiation

Based on the different temperatures of the building and its surroundings, we can distinguish between two opposite heat flow scenarios. In cold periods of the year when the average outdoor temperature is generally lower than the indoor temperature, the sum of all heat flows in a building is usually negative, mainly due to the energy output caused by transmission and ventilation heat losses (Fig. 2.2). In such cases, the ΔQ results in the amount of energy required for heating (Q_h—energy demand for heating) in order to reach a desired indoor temperature of approximately 20 °C, typical of cold periods.

The opposite is the warm period scenario, i.e., that of the summer period in the majority of European areas, when the highest daily outdoor temperature can be higher than the indoor temperature. The sum of all heat flows in a building results in a positive value, mainly due to solar heat gains. The ΔQ shows the amount of heat that has to be extracted from the building, (Q_c—energy demand for cooling), in order to reach a desired indoor temperature which should not exceed 25 °C. Energy flows described in this subsection refer to natural flows and do not take heat exchange caused by active technical systems into account.

Fig. 2.2 Energy flows in a building typical of cold periods

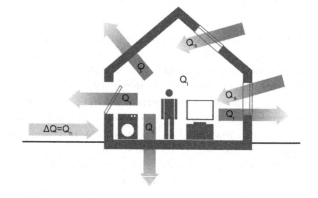

2.4 Climatic Influences and the Building Site

The importance of climate as a major determinant of the style of houses was pointed at as early as in Vitruvius [25]. Thorough inspection of the location has always been a first step in the process of planning and designing buildings. Numerous examples of vernacular architecture show how the building design responds to climatic conditions. Considerable differences in architecture typical of individual regions came to existence as a consequence of the response to specific location features comprising climatic characteristics of a larger region on the one hand and those of a particular location and its surrounding area on the other. From the point of view of bioclimatic planning, which is a basis for achieving the energy efficiency of buildings, location analysis is vitally important since numerous building designing aspects depend on the location specifics. It is thus possible to define the topography of the terrain, its soil composition and vegetation, the position and shape of the neighbouring buildings, the openness of the site, its orientation and most significantly, climatic circumstances. The latter have a major role in planning the building's heating, cooling and natural lighting strategies.

2.4.1 Global Climatic Impacts

Design principles considering the impact of the sun encompass two essential aspects: the apparent movement of the sun and solar radiation energy [24].

The earth moves around the sun in an elliptical orbit. At the same time, it spins in a counter-clockwise direction around its own axis once a day. The earth's axis is not normal to the plane of the earth's orbit, but tilted by 23.5°. Due to the tilt of the earth, not every place on earth gets an equal amount of sunlight every day, i.e., certain places have extremely short duration of daily light. As the earth revolves around the sun during a year, the angle between the earth's equatorial plane and the earth–sun line varies from +23.45° around 22 June to 0° around 21 March and 22 September, and to −23.45° around 22 December. This motion causes the phenomenon of seasons. For instance, when the northern pole is tilted towards the sun, the northern hemisphere experiences summer, while the places in the southern hemisphere get winter.

The plane of the earth's orbit is called the ecliptic and presents a reference plane for the positions of most solar system bodies. Since the earth orbits the sun, the sun is also on the ecliptic. Viewed from the earth, the sun appears to us as moving around the sky on the ecliptic. The apparent position of the sun can be determined by two angles, altitude and azimuth (Fig. 2.3).

The altitude (ALT) or solar elevation angle is the angle between the direction of the geometric centre of the sun's apparent disc and the idealized horizon, while the azimuth (AZI) is defined as the angle from due north in a clockwise direction. The angle indicating north is 0°, 90° for the east, 180° for the south and 270° for the west.

Fig. 2.3 Azimuth and
altitude angles for northern
latitudes

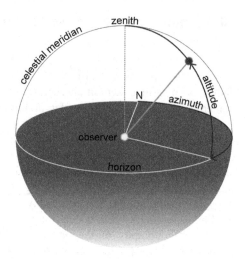

The apparent motion of the sun shows the sun as rising approximately in the east, moving through the south in a clockwise direction and setting approximately in the west. The sun rises due east and sets due west only on the first day of spring, 21 March and on the first day of autumn, 22 September. The apparent position of the sun varies for different hours of a day, days of the year and for different destinations.

For the purposes of energy-efficient building design, it is important to be aware of the sun's apparent movement when analysing the specified location of the building. Owing to the above-mentioned awareness combined with solar radiation and other climatic data, it is possible to make predictions for certain periods of the year and certain destinations in the sense of knowing where to lay focus in the process of designing, on passive solar heating or on prevention of overheating. Table 2.4 presents the position of the sun on two important dates, 21 June and 21 December.

The above data derived through free access to the sun-position-calculator software show divergence of the inclination angles of sunrays (ALT) at the summer solstice, from 54° in Tallinn, Estonia to 75.5° in Athens, Greece. In Tallinn, the sun rises at an azimuth angle of 36° east and sets at 324° west with the apparent sun path of 288°, which indicates very long summer days. On 21 December (winter solstice), the ALT at solar noon varies from 7.2° in Tallinn to 28.6° in Athens. The length of the day in Tallinn is very short, with the sunrise at an azimuth angle of 139° and the sunset at 221°, which shows that only southern façades can be directly exposed to the sun in winter. In Athens, the sun rises at an azimuth angle of 120° and sets at 240° (Fig. 2.4), with the apparent sun path of 120°, which indicates the longest winter day if compared to other cities from the Table 2.4.

Data for the main sun position over the course of a year have a crucial role in estimating elements such as the orientations of the building that will be exposed to

Table 2.4 Sun position at the summer and winter solstice for different destinations in Europe

Location	Latitude	Longitude	ALT solar noon on 21–06	AZI sunrise sunset on 21–06	ALT solar noon on 21–12	AZI sunrise sunset on 21–12
Tallinn	59.43°N	24.75°E	54°	36° 324°	7.2°	139° 221°
Copenhagen	55.68°N	12.56°E	57.8°	43° 317°	11°	133° 227°
London	51.50°N	0.12°E	61.9°	49° 311°	15.1°	128° 232°
Ljubljana	46.05°N	15.52°E	67.4°	54° 306°	20.6°	124° 236°
Madrid	40.42°N	3.70°E	73°	58° 302°	26.2°	121° 239°
Athens	37.98°N	23.73°E	75.5°	60° 300°	28.6°	120° 240°

Source http://www.timeanddate.com/ [27]

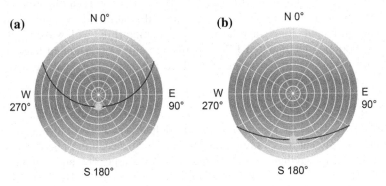

Fig. 2.4 Two-dimensional projection of the apparent sun path on: **a** 21 June and **b** 21 December, for Athens, Greece

direct radiation, the orientations of the glazing that will contribute to solar gains and the extent of the latter, the depth of the sunrays penetration into a room, the shape and size of the shading elements to be selected, etc.

For the effective implementation of passive solar design strategies, it is necessary to be aware of the basic facts about solar radiation. Approximately 70–75 % of the solar electromagnetic radiation enters the earth's atmosphere and reaches the earth. The electromagnetic solar radiation reaching the earth consists of three wavelength intervals (Fig. 2.5):

- Ultraviolet radiation (UV), 280–380 nm, which is harmful since it produces photochemical effects, bleaching, sunburn, etc.
- Visible light (VIS), 380–780 nm, ranging in colour from violet to red.
- Near infrared (NIR), 780–2,500 nm, also known as thermal radiation.

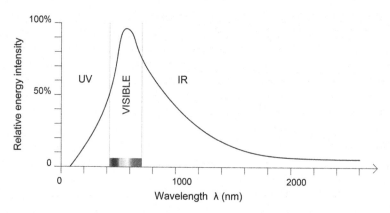

Fig. 2.5 Division of solar radiation according to the wave length

The incoming radiation is partly reflected back into space and partly absorbed by the atmosphere, clouds and the earth's surface. The absorbed radiation causes warming up of the earth's surface, and when the surface is warmer than the environment, it emits long-wave far-infrared radiation (FIR) with wave lengths of 2,500–50,000 nm. The emitted FIR is partly (15–30 %) transmitted back to the space and partly (70–85 %) reflected back to the earth. This leads to a further temperature increase on the earth.

Solar radiation can be measured as "irradiance" denoting the intensity [W/m²] or as "irradiation" denoting an energy quantity over a specified period of time [Wh/m²], [24]. There are large variations in irradiation at different locations on the earth due to different reasons, e.g., the angle of incidence, atmospheric depletion and the length of daylight period from sunrise to sunset. The annual total horizontal irradiation, also called global horizontal irradiance (GHI), for different destinations on the earth varies from approximately 400 kWh/m² near the poles to approximately 2,500 kWh/m² in the Sahara desert [24]. The global horizontal irradiance is the sum of the incident diffuse radiation and the direct normal irradiance (DNI) projected onto the horizontal surface, where the diffuse radiation is a combination of the radiation reflected from the surroundings and atmospherically scattered radiation.

Awareness of the solar radiation potential for a selected building site is vitally important for the purpose of achieving the energy efficiency. For instance, while planning to build a house in an area with low solar radiation and low average yearly temperatures, the main focus needs to be laid on excellent insulation. On the other hand, in the case of high solar radiation, the house should have larger south-oriented glazing areas, since winter solar radiation can be extremely beneficial for the building's energy balance. Data on solar radiation are treated as one of the main climatic indicators. In general, climatic conditions are one of the initial decision factors in designing an energy-efficient building.

2.4.2 Macro-, Meso- and Microclimate

Climatic conditions may be considered at three levels: macroclimate, mesoclimate and microclimate.

Macroclimate is a general climate of a region which encompasses large areas with fairly uniform climatic conditions. These vary from region to region due to the following factors: latitude (distance from the equator), altitude (height above sea level), topography (surface features), distance from large water bodies (oceans, lakes) and circulation of winds. Macroclimate is described by major climatic indicators provided by meteorological stations, such as temperature, humidity, air movement, i.e., wind (velocity and direction), precipitation, air pressure, solar radiation, sunshine duration and cloud cover.

Local characteristics of the area such as topography (valleys, mountains), large geometric obstructions, large-scale vegetation, ground cover, water bodies as well as occurrence of seasonal winds cause modification of general macroclimate conditions. These modified conditions denote the climate of a smaller area, also called mesoclimate. In [12], general types of mesoclimate having similar features are coastal regions, flat open country, woodlands, valleys, cities and mountainous regions.

The third level or the microclimate level is defined by taking into account human effect on the environment and consequently the way it modifies conditions within a specific area in the size of the building site. For instance, planted vegetation and nearby buildings influence the site's exposure to the sun and wind. Water and vegetation affect humidity whereas the built environment modifies air movement and air temperature [12].

Climatic classification systems define several climatic regions at the level of macroclimate. There is a variety of the existing climatic classification systems used for different purposes. One of the most recognized is the Köppen–Geiger system based on the concept of native vegetation. The original system underwent a number of modifications, which led to the current use of such modified systems [18, 22], distinguishing between five basic climate types; A-tropical, B-arid, C-temperate, D-cold and E-polar subdivided into subtypes according to temperature and precipitation data. Table 2.5 describes Köppen climate symbols.

Since the current book deals predominately with the building design suitable for European regions, Fig. 2.6 offers a more detailed description of European climate.

Europe is bounded by areas of strongly contrasting physical features that influence regional climate conditions. These areas are represented by the Atlantic Ocean to the west, the Arctic Sea to the north, a large continental part to the east and the Mediterranean Sea and north Africa to the south [12]. Northern zones influenced by north winds are known for cold winters with low solar radiation and mild summers. Mid-European areas close to the Atlantic Ocean influenced by humid western winds are known for cool winters and mild summers with a relatively high level of humidity reducing the strength of solar radiation. Central Europe has cold winters and warm summers, while southern Europe experiences

Table 2.5 Description of Köppen climate symbols and defining criteria [18, 22]

1st	2nd	3rd	Description	Criteria[1]
A			Tropical	$T_{cold} \geq 18$
	f		Rainforest	$P_{dry} \geq 60$
	m		Monsoon	Not (Af) and $P_{dry\_}100$–MAP/25
	w		Savannah	Not (Af) and P_{dry} <100–MAP/25
B			Arid	MAP <10°—$P_{threshold}$
	W		Desert	MAP <5°—$P_{threshold}$
	S		Steppe	MAP_5°—$P_{threshold}$
		h	Hot	MAT_18
		k	Cold	MAT <18
C			Temperate	T_{hot} >10 & 0 < T_{cold} <18
	s		Dry summer	P_{sdry} < 40 & P_{sdry} < P_{wwet}/3
	f		Dry sinter	P_{wdry} < P_{swet}/10
	w		Without dry season	Not (Cs) or (Cw)
		a	Hot summer	$T_{hot\_}22$
		b	Warm summer	Not (a) & $T_{mon}10\_4$
		c	Cold summer	Not (a or b) & $1\_T_{mon}10 < 4$
D			Cold	T_{hot} > 10 & $T_{cold\_}0$
	s		Dry summer	P_{sdry} < 40 & P_{sdry} < P_{wwet}/3
	w		Dry winter	P_{wdry} < P_{swet}/10 f—Without dry season Not (Ds) or (Dw)
	f		Without dry season	Not (Ds) or (Dw)
		a	Hot summer	$T_{hot\_}22$
		b	Warm summer	Not (a) & $T_{mon}10\_4$
		c	Cold summer	Not (a, b or d)
		d	Very cold winter	Not (a or b) & T_{cold} < −38
E			POLAR	T_{hot} < 10
	T		Tundra	T_{hot} > 0
	F		Frost	$T_{hot\_}0$

[1] *MAP* mean annual precipitation, *MAT* mean annual temperature, T_{hot} temperature of the hottest month, T_{cold} temperature of the coldest month, T_{mon} number of months where the temperature is above 10, P_{dry} precipitation of the driest month, P_{sdry} precipitation of the driest month in summer, P_{wdry} precipitation of the driest month in winter, P_{swet} precipitation of the wettest month in summer, P_{wwet} precipitation of the wettest month in winter, $P_{threshold}$ varies according to the following rules (if 70 % of MAP occurs in winter, then $P_{threshold}$ 2 × MAT; if 70 % of MAP occurs in summer, then $P_{threshold}$ 2 × MAT + 28, otherwise $P_{threshold}$ 2 × MAT + 14). Summer (winter) is defined as the warmer (cooler) six-month period of ONDJFM and AMJJAS

hot summers and mild winters with high solar radiation. A more accurate division of climate types according to the updated Köppen–Geiger climate classification is shown in Fig. 2.6. At the macrolevel, Europe is characterized by four climate types, where the dominant type according to the land area size is cold (D), followed by arid (B), temperate (C) and polar (E) climate types, with the latter being found within a smaller surface range [22]. All of the mentioned types are divided into subtypes of the second and third ranges (Table 2.5 and Fig. 2.6), which exhibit different features related to temperature and precipitation. Studies on the

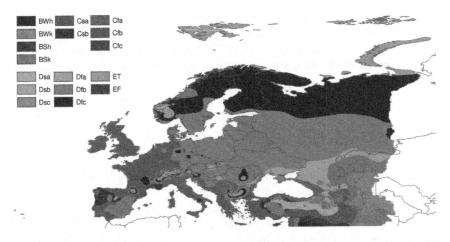

Fig. 2.6 European part of the updated world map of the Köppen–Geiger climate classification [22]

optimal glazing size and building shape presented in Sect. 4.3 are based on temperate and cold climates, Cfb and Dfb.

Apart from temperature and precipitation data, which are basic classification factors of the presented Köppen–Geiger system, the amount of solar radiation in combination with air movement characteristics composes another essential data base for the purposes of energy-efficient building design.

Solar radiation data can be obtained from maps of irradiation or through special software packages. The database is usually compiled from measurements of global horizontal solar irradiation and other meteorological and climatological parameters within a specific reference period.

Air movement affects thermal comfort of a building through convection and infiltration. Air movement or wind speed and its direction are usually measured at a height of 10 m. Wind data can be best presented by graphs with wind-rose diagrams showing the frequency of winds blowing from particular directions over a specific reference period.

When analysing a certain building site, it is necessary to consider climatic data at macro-, meso- and microclimate levels. As mentioned previously, the main climatic indicators can be modified to a certain extent by local topography, vegetation, surrounding buildings, etc. For instance, daily air temperature in wooden areas can be lower by a few degrees than that of open areas, since tree foliage reduces the amount of solar radiation hitting the ground. In the nighttime, tree foliage impedes the outgoing long-wave radiation and a drop in the air temperature is therefore lower [12]. The air temperature can also be influenced by topography, ground surface, the surrounding buildings and the vicinity of water areas. Likewise, solar radiation can be weakened by dust particles in the air or largely hindered by some geometric obstructions like hills, buildings or, as explained beforehand, by vegetation.

While the general macro- and mesoclimates of the region are beyond human influence, some significant benefits can be provided by human effect on the environment at the microclimatic level [12]. All in all, a good understanding of the regional and local climate means an essential input into a most effective building design process.

2.5 Basic Design Parameters

As mentioned beforehand, a complex design approach to energy-efficient buildings may be conducted at three levels. The primary or basic design level includes determination of the building shape and orientation in addition to the arrangement of interior spaces and selection of the building components. Energy balance of the building is enhanced through design of passive strategies, while the final design phase step goes to application of active technical systems.

2.5.1 Building Shape

The building shape is defined by geometry of external building elements, such as the walls, floor slab and roof. It has a significant effect on thermal performance, since major heat flows (cf. Sect. 2.3) pass through the building envelope.

The building shape can be expressed by the shape factor (F_s) defined by the equation:

$$F_S = \frac{A}{V} \; [\text{m}^{-1}] \tag{2.2}$$

where the equation quantities are the following:
A   total area of the building thermal envelope [m^2]
V   total heated volume of the building [m^3]

To minimize transmission losses through the building envelope, a compact shape indicated by a low shape factor is desirable (Fig. 2.7). On the other hand, a dynamic form with larger transparent surfaces enables provision of higher solar gains. Integrating the aspect of solar access into the phase of determining the building shape is an essential part of energy-efficient building design [13]. According to general design guidelines for energy-efficient houses, a compact rectangular shape is seen as the optimum. However, some studies show that under certain location and climatic conditions, a dynamic shape might be even more efficient as far as energy gains are concerned. Further interesting findings relating to the building shape are presented in Sect. 4.2.2.

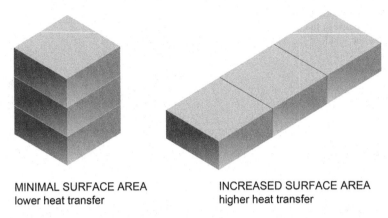

MINIMAL SURFACE AREA   INCREASED SURFACE AREA
lower heat transfer   higher heat transfer

Fig. 2.7 Shape factor is defined by the ratio between the area of the building thermal envelope and the heated volume

Another parameter referring to the proportions of the building is called the building aspect ratio (AR). It is defined as the relationship between linear dimensions of individual external building elements. It can denote the relationship between the building's height and width ($AR_H = H/W$) or between the building's length and width ($AR_L = L/W$). When considering solar access to selected building elevations, the relationship between the equator-facing façade length and the lateral façade length can play an important role. Design guidelines usually require that buildings be oriented with a longer glazed façade to the south to ensure higher solar gains. Figure 2.8 presents a relationship between the length and the width of the building.

When buildings feature more dynamic and fractured forms, the issue of solar access to individual parts of the envelope proves to be of great significance since some parts of the building could shade the others. Many studies therefore treat the geometric relationship between the shaded and exposed part of the building envelope and its influence on solar gains. Figure 2.9 presents a relationship between the exposed and partly shaded façade.

Defining the building shape should respond to the building's environment and particularly to climatic conditions. Observation of vernacular buildings in different climatic regions shows a number of differences. Massive walls, small windows and flat roofs are typical of buildings in hot and dry climates with significant diurnal temperature differentials. Massive walls ensure adequate thermal mass for thermal inertia (cf. Sect. 2.6.1), small windows prevent overheating during summer days and flat roofs prove to be a suitable choice, since there is usually little precipitation. Even the principles of spatial planning take high solar radiation into account, which leads to buildings often being closely clustered for the purposes of shading one another and the public spaces between them. On the other hand, vernacular building design is quite different in hot and humid climates. The use of large windows exposed to cooling breezes enables summer cooling while large

Fig. 2.8 Aspect ratio L/W—relationship between the length and the width of the building

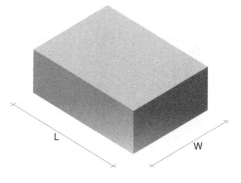

Fig. 2.9 Geometric relationship between the exposed and partly shaded façade

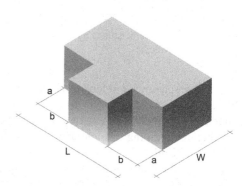

overhangs and shutters provide protection from solar radiation and rainfalls [7]. Massive structure is not typical since there are small differences between day—and nighttime temperatures. Finally, buildings in a predominantly cold climate need a compact form with a minimal shape factor reducing heat losses through the envelope.

2.5.2 Orientation

Building orientation is defined as the angle between the normal to a certain surface, e.g., façade, and the north cardinal direction. It is determined for each of the building's façades. North orientation is thus defined by the angle of 0°, south by the angle of 180°, east by the angle of 90° and west by the angle of 270°. An additional explanation regarding the terms "south, north, east and west" is necessary at this point. "Facing the equator" or "equatorial orientation" are terms frequently used to describe the orientation with respect to the northern and southern hemispheres. For the northern hemisphere, equatorial orientation denotes south orientation. Since the current book and especially the studies presented in

Sect. 4.3 deal mainly with buildings located in the northern hemisphere, the term "south" orientation will be used as a synonym for equatorial orientation.

With respect to guidelines for energy-efficient housing, a major part of the building's transparent surfaces should be oriented to the south. South orientation enables higher solar gains and better daylighting, but at the same time, it increases a risk of summer overheating. In order to prevent overheating, a well-designed solar control is indispensable for the buildings located in a great number of European regions. A study treating the issue of the optimal size and distribution of the glazing surfaces can be found in Sect. 4.3.1.

Certain building sites may be less favourable in terms of orientation, since they cannot enable the orientation of the building mainly to the south. The task of architects in such cases is to take maximum advantage of the existing conditions by adjusting the design concept to suit the given microlocation.

An exemplary case of such design pursuit is the Sunlighthouse in Pressbaum, Austria, developed by the architects of Hein-Troy Architekten for the Model Home 2020, a project set by the Velux Group. The specifics of the terrain and the vicinity of the neighbouring buildings in addition to the mainly south-east-oriented site were a challenge for the architects who conceived well thought out strategies, such as positioning of the atrium, using different roof slants, allowing for natural lighting to penetrate through the roof windows, in order to prove that a house can have an excellent energy certificate and a comfortable indoor climate in spite of the rather undesirable microlocation conditions (Fig. 2.10).

2.5.3 Zoning of Interior Spaces

Zoning is a term used for division of a building into spaces having similar characteristics based either on the purpose of individual spaces or on interior climate conditions. Energy-efficient building design requires careful consideration of indoor zoning in order to create a rational distribution of heat and daylight. There are several zoning concepts which strongly depend on the type, size and the purpose of a building.

Appropriate thermal zoning can result in lower heat flow between individual spaces in the building or between the building and its surroundings. One of the frequently used thermal zoning concepts suggests positioning rooms with a low interior temperature (e.g., staircase, pantry and entrance hall) next to the north-facing exterior wall which tends to be cooler than the south-facing wall, due to less intense exposure to solar radiation. On the contrary, spaces such as the living room, dining room or bedrooms for children which demand a slightly higher air temperature belong to the south-facing side where additional heating is provided through solar radiation (Fig. 2.11). Another important part of building design is vertical thermal hierarchy; a basement placed inside the thermal building envelope needs to have good insulation and constant heating while a non-heated basement should be placed outside the thermal envelope. In the latter case, the entire ground

Fig. 2.10 Sunlighthouse in Pressbaum, Austria (Photo by Samo Lorber)

floor above the basement is required to have suitable insulation. The staircase, a special zone and a connecting element between a cold basement and the heated living area, thus becomes a critical point of thermal bridging. In order to prevent heat losses, a buffer zone with insulating doors must be planned, which leads to a thermal bridge being avoided.

Thermal zoning in public buildings tends to be more complex since a basic linear alignment of spaces to the north, middle and south zones does not always prove to be suitable. Nevertheless, it is sensible to group rooms with similar characteristics even in large-size buildings for the purposes of functionality and heating requirements.

A further essential element to be foreseen is daylight zoning. Rooms functioning as places for work, play or residence require substantial lighting and are thus positioned alongside the glazed façades while a more central placement suits rooms demanding less daylighting. The result of such zoning is excellent indoor climate and lower electricity consumption.

Fig. 2.11 Thermal zoning
of interior rooms in a
single-family house

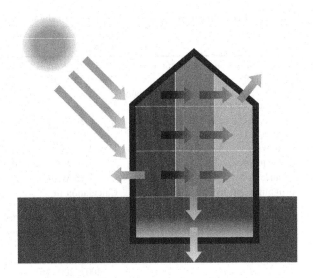

2.5.4 Building Components

In general, a building consists of structural elements, elements of protection and finalization and active technical systems. Structural elements can be made of different building materials, such as brick, concrete and timber. A most important role in protecting the building against weather conditions, heat, vapour or noise is attributed to insulation materials whose proper selection partly depends on the type of structural system. All building components should be fully coordinated with the building's requirements determined by its purpose, its occupants and climatic conditions. The term "thermal envelope" refers to building components embracing the heated volume of a building creating a barrier, which prevents unwanted heat exchange between the interior and the exterior of the building. The efficiency of the thermal envelope depends on material composition of the walls, roof and floor, on the type and size of transparent surfaces, on solar control, air tightness, thermal bridges, etc.

In order to achieve a certain level of energy efficiency, a range of specific requirements, among which some are related to building thermal envelope, are prescribed in national building codes or standards on the basis of certificated labels widely used in construction industry. A comparison of the requirements set in the national building codes of Germany [8] and those set by Passive House Institute [10] are presented in Table 2.6.

Substantial deviations between the listed measures for the quality of the building envelope can be observed from the table above. A building can be treated as energy-efficient when its design satisfies all the requirements set in the building codes. Other standards based on certified labels, such as Minergie and Passive House, determine more stringent criteria; houses built by these criteria therefore exhibit a much better energy performance.

Table 2.6 Requirements related to the building thermal envelope set in different building codes

Building component	Reference values on the basis of EnEV [8]	Reference values on the basis of passive house [10]
Exterior walls against ambient air	$U = 0.28$ W/m^{2}K	$U = 0.15$ W/m^{2}K
Exterior walls and floor slabs adjacent to the ground, walls and floor slabs against unheated space	$U = 0.35$ W/m^{2}K	$U = 0.15$ W/m^{2}K
Roof, top floor ceiling, knee walls in attics	$U = 0.20$ W/m^{2}K	$U = 0.15$ W/m^{2}K
Windows	$U = 1.30$ W/m^{2}kg ≥ 0.60	$U = 0.80$ W/m^{2}kg ≥ 0.50
Roof windows	$U = 1.40$ W/m^{2}kg ≥ 0.60	$U = 0.80$ W/m^{2}K ≥ 0.50
Entrance door	$U = 1.80$ W/m^{2}K	$U = 0.80$ W/m^{2}K
Airtightness of the building envelope at a pressure differential of 50 Pa	Calculation in accordance with DIN V 4108–6: 2003–2006 [5]; with air leakage testing DIN V 18599–2: 2007–2002 [6]; by class I	0.6

The following subsections contain presentation of the components essential for the quality of the building thermal envelope.

2.5.4.1 Structural Materials and Construction Technology

Given a wide range of properties indicating applicability of different building materials for energy-efficient construction, the selection of the most suitable construction technology strongly depends on several factors. Among these, the most crucial are the building location, climatic conditions, availability of construction technologies and materials. With respect to climatic conditions, massive construction is more appropriate than lightweight construction for hot and dry climates, since thermal mass plays an important role in prevention of overheating. It could be claimed that timber construction is most appropriate for the majority of European climates, either cold or temperate with the exception of regions with hot summers. The selection is based on a combination of criteria taking into account environmental impact (life cycle analysis, mode of material production and its embodied energy), speed of construction, physical, mechanical and structural properties, potentials on recycling and reuse, etc. Timber construction distinguishes between six main structural systems and a few subsystems, all fully described in Chap. 3.

Furthermore, the use of glazing surfaces in timber structures is becoming an important issue of energy-efficient construction. Over years of development, glazing manufacturers have in fact improved their products' thermal insulation and strength properties as well as their coefficient of permeability of the total solar radiation and thus enabled the use of large glazing surfaces, primarily south-oriented, not only to illuminate indoor areas but also to ensure solar heating. It follows that such timber-glass structures represent a great potential in residential and public building construction.

2.5.4.2 Insulation Materials

The selection of the type and quantity of insulation installed in building elements depends on the climate, on the type of the construction element that has to be insulated, on construction technology and on general building properties.

The purpose of application is the criterion for the four basic groups of insulation materials providing:

- Thermal protection
- Sound protection
- Protection from moisture
- Protection from air leaking.

Certain insulating materials feature two functions at the same time, e.g., thermal and sound protection. Among the above-listed groups that of thermal insulation, i.e., protection from unwanted air infiltration and moisture has the highest impact on energy behaviour of buildings, while a comfortable living climate also depends on proper application of sound-insulating materials.

Thermal Insulation

Thermal insulation reduces the average heat flow through elements of the building thermal envelope. The thermal efficiency of insulating materials is expressed by the coefficient of thermal conductivity (λ) or by the coefficient of thermal transmittance (U) for composite elements. U-value signifies the heat flow through 1 m^2 of the wall area at a constant temperature difference of 1 K. Sufficient insulation helps to reduce the conductive heat gains into a building in the summer period and limits the conductive heat losses when the external temperature is lower than the interior temperature.

Many different materials are used for the purposes of thermal insulation. Considering their chemical composition, physical structure and resource insulating materials, they can be classified into inorganic, organic and new innovative materials presented in Table 2.7.

A range of different innovative materials (column 4, Table 2.7) are currently under development and their use is expected to increase in the future. Their biggest advantage in comparison to traditional insulating materials is smaller thickness

Table 2.7 Classification of common insulating materials, adopted from Papadopoulos [21] and Jelle et al. [16]

Inorganic	Organic	Combined	Innovative
Foamy	**Foamy**	Calcium silicate	Transparent insulation
(mineral	Expanded polystyrene	Gypsum	Aerogels
source)	Extruded polystyrene	foam	Vacuum insulation panels (VIP)
Foam glass	Polyurethane foam	Wood wool	Gas-filled insulation panels
Expanded			(GFP)
clay			*Phase-change materials (PCM)*
Vermiculite			*Thermochemical materials (TCM)*
Perlite			Dynamic materials
Fibrous	**Expanded foamy**		
(mineral	Cork (plant origin)		
fibres)	Phenolic foam		
Glass wool	Melamine foam		
Stone wool			
	Fibrous (natural		
	fibres)		
	Sheep wool		
	Cotton wool		
	Flax		
	Cellulose		
	Coconut fibres		
	Wood fibres		
	Straw		
	Hemp		

that can provide even better insulation features. Although not being typical insulating materials, PCM and TCM are mentioned due to their role in thermal storage. More information on innovative insulating materials and those still developing can be found in Jelle et al. [16] and Jelle [17].

Traditional insulating materials (listed in columns 1–3, Table 2.7) feature similar performance in terms of thermal conductivity whose value ranges from approximately 0.025–0.050 W/mK. On the contrary, these materials are characterized by significant differences related to their properties, such as

- Physical and mechanical properties (density, mechanical strength, fire resistance, moisture resistance, sound absorption)
- Environmental impact (primary embodied energy, carbon footprint, ability for reuse and recycling, use of additives against biological impacts, classification of their treatment as waste, etc.)
- Health aspect (dust emissions, toxicity during fire, level of health hazardous substances, etc.)
- Cost
- Area of application.

A detailed list of insulating material properties can be found in Papadopoulos [21] and Jelle [17]. Owing to the possibility of being manufactured in different densities even a single type of insulating material can have different physical properties, such as thermal conductivity, sound resistance, strength characteristics, thermal inertia. Figure 2.12 offers a graphical presentation of some common insulating materials mostly applied in the construction of energy-efficient buildings.

Besides thermal conductivity and density, the specific heat capacity (c) of each material composing the building envelope is crucial for the dynamic response of a building. These properties influence the time interval, also called *time lag*, between

foam glass granulate   foam glass   expanded clay

glass wool   stone wool   cellulose

expanded polystyrene   extruded polystyrene   cork

wood fibre boards   straw   flax

sheep wool   coconut fibres   hemp

Fig. 2.12 Graphical presentation of different insulating materials

the moment when the highest temperature appears on the external building's surface and the moment when the highest temperature is reached on the internal surface. In practice, the term *phase shift*, measured in minutes per centimetre of the material, is commonly used to describe how fast the heat will progress through the material [min/cm]. With elements composed of more materials, the phase shift can be recalculated to express an overall temperature time lag. In order to store more heat and achieve a longer time lag, materials composing external building elements need to have high heat storage capacity and high density. The recommended value for the time lag in most of European climatic regions is around 12 h or slightly more. This property reflecting thermal stability of buildings is particularly important in summer to prevent overheating. For example, if the highest temperature appears on the external building's surface at 2 pm in summer, the highest temperature of the interior appears in 12 h, at 2 am when the external temperatures are usually lower and the cooling of the interior can be thus performed through natural window ventilation.

A comparison of different insulating materials (Table 2.8) shows excellent properties of cellulose which has low thermal conductivity ($\lambda = 0.035$–0.040 W/mK), high heat capacity ($c = 1900$ J/kgK) but relatively low density (30 to 60 kg/m^3) in comparison to other standard insulating materials. Wood-fibre boards have slightly higher thermal conductivity ($\lambda = 0.045$–0.055 W/m^{2}K) but even higher heat capacity ($c = 2100$ J/kgK). Along with their density (190 do 270 kg/m^3), wood-fibre boards achieve up to 6 times longer phase shifts than other materials.

Due to the different heat storage capacities, it is also sensible to use different types of insulation materials in a single building. In order to achieve a longer phase shift, insulation materials of higher density and higher heat storage capacity are recommended for the protection of the parts of the building envelope that are extremely exposed to solar radiation, such as south-oriented roofs and façades.

Another important property of insulating materials is resistance to vapour diffusion (μ) which needs to be as low as possible. The latter characteristic along with the thickness of the material provides data on the resistance of vapour passage through a certain material [Sd in metres]. *Sd* value reveals the air layer thickness, i.e., the resistance of the air layer through which water vapour needs to permeate,

Table 2.8 Physical and mechanical properties of different insulating materials; adopted from Papadopoulos [21]

Insulating material	Thermal conductivity λ [W/mK]	Density ρ [kg/m^3]	Heat capacity c [J/kgK]	Phase shift [min/cm]	Resistance to vapour diffusion μ
Glass wool	0.030–0.045	13–100	840	7–10	1
Stone wool	0.033–0.045	30–180	840	10–19	1
EPS*	0.029–0.041	18–50	1,260	8–15	25–60
Wood fibres	0.045–0.550	190–270	2,100	42–50	10
Cellulose	0.033–0.040	30–60	1,900	18–28	1

* Expanded polystyrene

e.g., $Sd = 1$ m is equivalent to the resistance of the air layer with a thickness of 1 m. Organic foamy materials, expanded polystyrene, extruded polystyrene and polyurethane foam have relatively high resistance to vapour diffusion. In the case of diffusion-open construction that has been currently gaining in popularity, such insulating materials can either not let water vapour penetrate at all or allow for its passage only to an extremely low degree. As a consequence, condensation layers reducing the insulation capacity of the materials may appear on their surface. Using these materials to insulate timber structures means a risk of excessive condensation where humidity is forced into wood, thus leading to the growth of fungi which destroy both, wood and the insulation material. On the other hand, insulating timber structures with low-diffusion resistance materials permit a natural passage of water vapour through the building envelope. Another advantage of thermal insulating materials mainly based on wood (wood-fibres boards, cellulose) is seen in water vapour being temporarily—until the drying out is reached—distributed within a fairly large insulation area, which does not significantly alter thermal insulation properties.

For the purposes of energy-efficient building design most appropriate insulating material should therefore feature low thermal conductivity, high thermal storage capacity, relatively high density and low resistance to vapour diffusion. Such materials can be applied to above-the-ground construction elements, such as roof and walls. In the summer period, the roofs are more exposed to solar radiation than the walls, mainly due to their inclination angle and to the absence of shading elements in their surroundings. Therefore, proper roof insulation with a high phase shift is vitally important in prevention of overheating. Owing to smaller differences between the internal and ground temperatures, the phase shift is not as important for construction elements adjacent to the ground, such as a slab on grade, a basement slab and wall, whose thermal conductivity values need not exceed the average. The insulating material installed in a slab on grade, a basement slab or a basement wall is required to have relatively low thermal conductivity, but high density and high resistance to humidity and vapour diffusion. Materials with such properties are extruded polystyrene, foam glass granulate and expanded clay aggregate.

Apart from the physical and mechanical properties, environmental and health aspect should also be considered in the process of sustainable building design. Table 2.9 presents the main environmental factors of different insulating materials.

The importance of the properties described in this section is relevant not only to insulating materials but also to all the other building materials installed in a building. Since the current book treats timber-glass buildings with two major building materials, timber and glass being already predefined, the emphasis relating to the selection of appropriate building materials should be directed mainly to insulating materials, with a special focus laid on sheathings, plastering, foils, etc. In combination with timber and glass, these materials have to display optimal performance in order to achieve a high level of the building's energy efficiency and a comfortable indoor climate.

Table 2.9 Environmental properties of different insulating materials adopted from Papadopoulos [21] and Hammond and Jones [14]

Insulating material[*]	Primary embodied energy [MJ/kg]	Primary embodied energy [MJ/kg]	Embodied Carbon[***] [kg CO_2/ kg]	Reuse and recycle	Waste disposal
Glass wool	28–41	28.0–40.0	1.35	Not reusable not recyclable	No limitations
Stone wool	16	16.8	1.05	Not reusable not recyclable	No limitations
EPS[**]	89–105	109.2	3.40	Reusable recyclable	Long bio persistence as a waste material
Wood fibres	20	10.8–20.0	to 0.98	Reusable recyclable	Biodegradable in landfill
Cellulose	3.5	3.3	−2.00	Highly recyclable	

[*] Minor deviations in values can appear due to different implementations (density) of individual insulating materials and to variable data sources
[**] Expanded polystyrene
[***] Embodied carbon (cradle-to-grave)

2.5.4.3 Glazing Surfaces

As a transparent component of the building skin, glazing surfaces can be found in windows and doors or as glass roofs and glass façades. Providing daylight and ensuring visual contact between the interior and the exterior are the main functions of the glazing, in addition to weather protection, thermal protection, noise protection and natural ventilation in the case of operable windows.

Within the concept of energy-efficient building, glazing surfaces have to be designed to suit the needs of saving energy. The heating and cooling energy requirements of a building are largely determined by heat losses and gains through the building envelope which consists of transparent and non-transparent building elements. Since the average U-value of the glazing is usually higher than that of other non-transparent building elements, glazing surfaces present a potential risk of the highest heat transmission losses. On the other hand, glazing areas enable solar radiation to enter the building, which is a basis for the passive solar building design. The amount of solar gains has to be controlled to prevent overheating in warm periods of the year, and an integrated shading design is therefore necessary. Only the contemporary insulating glass units can assure appropriate energy exchange satisfying the requirements of energy-efficient building design.

In order to understand the principal glazing functions and compare different glazing structures, knowledge of material properties of glass and understanding of the basics of building physics is required. These topics are presented in Sect. 4.2.

Moreover, the importance of the glazing orientation and its size is analysed in the studies contained in Sect. 4.3.1.

2.5.4.4 Solar Control

Solar control strategy design calls for identification of a situation in order to establish whether the solar input is desired or needs to be excluded. Solar control requirements mainly depend on the climate, time of the year and the characteristics of the location. For instance, the winter period when solar penetration into buildings is desired and the summer period when it should be reduced to avoid overheating are two diametrically opposite situations typical of moderate climates. Measures suitable for cold climates where solar radiation is more or less desired differ from those applicable in the case of hot climates where the excess solar radiation entering a building might increase the cooling load and make the prevention of direct solar heat gain necessary for a longer period of the year. The time of the year along with the orientation of the exposed windows requires a specific shading approach. While shading design of windows facing the equator (southward in the northern hemisphere and northward in the southern hemisphere) tends to be simple, it is far more complicated to shade eastern and southern façades. There is almost no need for shading north-oriented windows whose exposure to direct solar radiation is almost nonexistent, with the exception of early summer morning and late evening sun. Effective solar control can also prevent uncomfortable glare which can appear in any of the above-mentioned situations. Solar rays at low altitudes are likely to cause even more glare problems than the heat gain.

In order to develop a proper shading strategy, it is essential to understand the basics of solar geometry (cf. Sect. 2.4.1). The apparent movement of the sun during a day and during a year calls for a complex shading approach which can be best achieved with a combination of different strategies. Sources of solar radiation requiring shading are direct radiation and radiation reflected from the surroundings (Fig. 2.13).

Solar penetration through glazing surfaces can be reduced by means of different measures, such as shading devices among which only the external ones prove to be effective, landscaping or also glass coatings [2]. An important role within a shading strategy is furthermore played by vegetation, the surrounding buildings, structures, hills and slopes. Each of the listed measures has its advantages and is more or less appropriate for one of the orientations.

Shading with vegetation to protect low-rise buildings is effective only if broad-leaved trees are planted, since they block the sun in the summer and let it through their bare crowns in wintertime. A further important element is seen in the positioning and the height of the crowns. If shading is desired on the eastern and western building façades, where the solar incidence angle is always low, the trees should have low crowns and be placed at a certain distance from the building (Fig. 2.14).

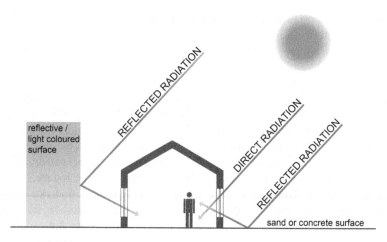

Fig. 2.13 Sources of solar radiation requiring shading

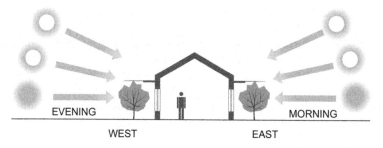

HORIZONTAL OVERHANGS ARE INEFFICIENT ON EASTERN AND WESTERN FAÇADES

Fig. 2.14 Shading requirements for eastern and western façades

On the contrary, the best effect for the south-oriented façade is achieved if the latter is shaded with broad-leaved trees having high crowns (Fig. 2.15). Growing vines combined with pergolas can also be used for shading purposes. An additional benefit of solar control provided by vegetation is its contribution to a better microclimate while on the other hand, it is not advisable to rely solely on this strategy due to a risk of unexpected growth of a tree or its contracting a disease [2].

In addition to solar protection of the glazing surfaces, shading of pavements surrounding the building by means of shrubbery can help reduce heat absorption and subsequent heat radiation to the building and its surrounding area [2].

More reliable shading measures can be taken through external shading devices. These are more effective at reducing unwanted solar gain than internal shades and stop the sun from penetrating through the glazing. External shading devices can be fixed or operable. Fixed devices are usually cheaper but do not allow for optimal year-round adjustments to the geometry of the solar inclination angles.

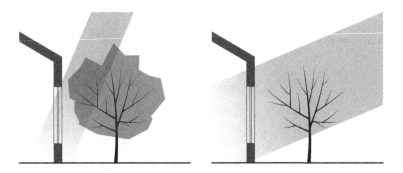

Fig. 2.15 Shading provided by vegetation

General types of external shading devices are vertical, horizontal and egg-crate devices [24], in addition to screens.

Horizontal shading elements are most suitable for southern exposures (Fig. 2.16). They are intended to reduce solar radiation of high inclination angles and are most effective when the sun position is nearly opposite to the shaded window. Best results are achieved if shading devices are installed externally, while in the case of double-skin façades, the shading elements, mostly roll-down blinds, are usually installed in the interspace between the first and the second skin.

Typical horizontal elements are roll-down blinds, shutters, louvres and overhangs. The latter can also be designed as canopies, pergolas and strips. Overhangs must be carefully planned in order to let the sun rays enter the building in winter and to reduce solar penetration in the summer period when the sun's angle of incidence is the highest (Fig. 2.17) [3].

Overhangs in eastern or western exposures need to be extremely deep to achieve the shading effect, and their use in these orientations is therefore not advisable unless a deep overhang can be used as a canopy providing a comfortable, usable outdoor space for the occupants. An adequate depth and height of the overhang can be determined with a simple calculation method which can be found at http://windows.lbl.gov/daylighting/designguide/section5.pdf [28]. Since the year-round need for shading is of a changeable nature, the use of movable shading devices proves to be a practical solution. Awnings, for example, whose function is to provide protection in the summer can be adjusted or totally folded down in the winter season.

Vertical shading devices (Fig. 2.18), such as louvers or projecting fins, are most suitable for eastern and western façades exposed to solar radiation of lower inclination angles reaching the façade from the south-eastern or south-western direction. Vertical devices also block the early morning sun coming from the north-east or the late evening sun coming from the north-west direction, which a summer period phenomenon typical of certain geographic regions (see Table 2.4). In fact, with the early morning sun of lower intensity, this type of shading serves rather as prevention from glare than from overheating.

Fig. 2.16 Basic types of
horizontal shading devices
for southern exposures

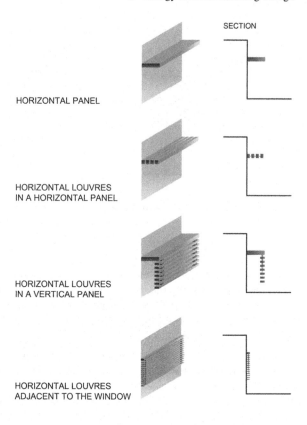

SECTION

HORIZONTAL PANEL

HORIZONTAL LOUVRES
IN A HORIZONTAL PANEL

HORIZONTAL LOUVRES
IN A VERTICAL PANEL

HORIZONTAL LOUVRES
ADJACENT TO THE WINDOW

Egg-crate devices (Fig. 2.19), like grilles made from metal, concrete or wood, are most effective for orientations which are in-between the east and south or the east and west, since they block sun rays of different altitude and azimuth angles.

Screens (Fig. 2.20) reduce solar radiation independently from the inclination angle. They can be opaque or semitransparent. A disadvantage of the screens is blocking of the view to the outside, which makes this kind of shading suitable when maximum dimming of the interior is required. Since not primarily intended for the prevention of overheating screens can be also installed on the internal side of the window.

2.5.4.5 Air Tightness

Leaking of the building envelope results in a number of problems most often arising in the winter period when cold air infiltration increases the energy demand for heating. A minor leakage problem used to be tolerated with the explanation that it contributes to fresh air exchange although the truth is that it cannot provide a sufficient quantity of fresh air and it can consequently not replace the necessary ventilation.

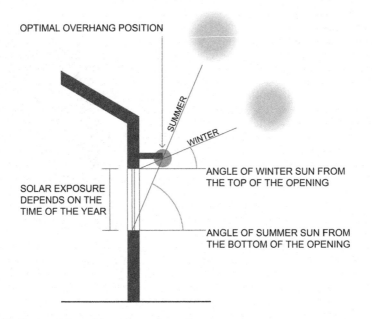

OPTIMAL OVERHANG POSITION

SUMMER

WINTER

ANGLE OF WINTER SUN FROM
THE TOP OF THE OPENING

SOLAR EXPOSURE
DEPENDS ON THE
TIME OF THE YEAR

ANGLE OF SUMMER SUN FROM
THE BOTTOM OF THE OPENING

Fig. 2.17 Design strategy for overhangs in southern exposures

Fig. 2.18 Basic types of vertical shading devices for eastern and western exposures

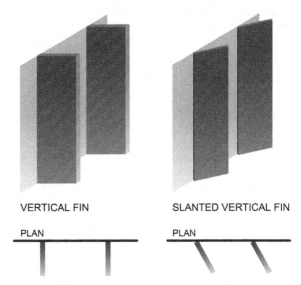

VERTICAL FIN

SLANTED VERTICAL FIN

PLAN

PLAN

With regard to the requirements for energy-efficient house design, the level of air tightness differs from one energy class to another. In the case of a passive house standard, where mechanical heat recovery ventilation is required, the air tightness of a building has to be excellent. The air leakage must not exceed 0.6 h^{-1} (volume per hour) at a pressure

Fig. 2.19 Egg-crate shading
devices for exposures
declined from the south

EGG CRATE

PLAN

Fig. 2.20 Screen shading
devices intended for the
prevention of daylight in the
interior

SCREEN SHADING

PLAN

differential of 50 Pa measured by the so-called blower-door test. In houses with no mechanical ventilation system, the air tightness can be accordingly smaller. The air tightness is achieved using an airtight layer running inside the thermal envelope of a building. The appropriate materials to use for an airtight layer are internal plaster, plywood board, particle board, OSB, durability-stabilized plastic foils, bituminous felt and finally, a frequently used tear-proof building paper. Special attention has to be paid to the building element junctions, where a certain degree of air tightness has to be assured as well.

2.5.4.6 Thermal Bridge-Free Construction

Thermal bridges in building envelopes are caused by the occurrence of zones with relatively high thermal conductivity creating pathways for heat losses. Usually, these zones arise from the lack of insulation or use of highly conductive building materials or even due the specific building geometry.

Energy-efficient house design demands a reduction in thermal bridges to a minimum, which can be generally achieved by making a continuous thermal envelope with no interruption of the insulating layer (Fig. 2.21).

Windows are an exception since the thermal envelope is interrupted at the point of their placement. Nevertheless, thermal bridges can be maximally reduced if a highly insulated window is installed correctly with an airtight multiple sealing. Frequently, the most difficult thermal bridges to avoid are those of below-the-ground building elements which are affected by the soil and must be treated differently than thermal bridges to the ambient air [10].

The use of prefabricated timber-frame structural elements is a highly suitable approach in avoiding thermal bridging, which leads to reduction in heat losses and prevention of building damages.

2.6 Design of Passive Strategies

As the basic design parameters are finally determined, the second level of energy-efficient design commences. Passive design strategies allow for the passive use of natural energy sources and climatic indicators in order to ensure an adequate indoor climate and reduce the need for active heating, cooling, ventilation and

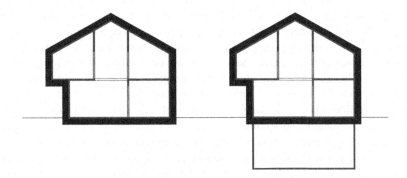

Fig. 2.21 Thermal envelope not interrupted with thermal bridges

daylighting. For example, solar radiation can be exploited to assist in reducing energy consumption for space heating and daylighting, natural ventilation can assist in reducing energy consumption for mechanical ventilation and cooling, etc.

Application of specific strategies largely depends on climatic conditions, type and occupancy of the building. The general principle is to maximize free heat gains which have to be equally distributed and stored within the building in periods with lower average outdoor temperatures and to minimize heat gains and assure natural cooling through ventilation in warm seasons. The exploitation of daylight should be part of the year-round strategy.

2.6.1 Passive Heating Strategy

The majority of European regions prove to be suitable areas for passive solar design. The key to successful passive solar design is in respecting the character-istics of the building site and its solar exposure, local landscaping, regional climate and microclimatic specifics. Solar energy can be used for heating, daylighting and water heating. While the heating of domestic water requires utilization of the equipment consuming electricity, solar heating and daylighting of interior spaces need no help of active technical systems, hence the name passive strategies.

A complex passive heating strategy integrates several approaches, e.g., solar collection, heat storage and conservation along with heat distribution [12], in addition to solar control aimed at preventing the overload of solar energy.

Solar collection in the building is essential for the passive heating strategy. The quantity of solar gain depends on the properties of elements hit by sunrays, the area and inclination of these elements, the angle of solar incidence and the available solar radiation. Furthermore, orientation, topography and shading influ-ence the amount of solar radiation reaching the building envelope and entering the building. A number of studies have been performed with the objective to analyse the most suitable orientation, glazing size, quality and inclination angle for the best exploitation of solar radiation. A common finding of all these studies carried out for different climatic conditions is seen in their definition of southern orientation as the most effective position for solar collection. The optimal share of south-oriented glazing depends on climate and thermal properties of the building envelope. A research related to the optimal glazing size will be explained in Sect. 4.3. North-oriented glazing receives direct solar radiation only in summer, with the early morning and late evening sun, which means no gains are available in winter time when solar passive heating is most desired. East- and west-oriented windows receive a similar amount of solar radiation, with the west-facing windows con-tributing to overheating in summer in the case of inadequate shading. The incli-nation of glazing surfaces plays a further vital role relevant to the amount of solar gains. The highest transmittance of solar radiation appears when the solar beams hit the glazing perpendicular to its plane. During the summer period, the sun's altitude at solar noon ranges between 54° and 75° in most European cities and the beams hit vertical glazing at a sharp angle, which results in lower transmission and consequently in lower solar gains. Higher transmission can occur through tilted glazing installed in pitched roofs where solar beams hit the glazing plane at angles close to the perpendicular to the glazing plane. Tilted glazing therefore transfers large amounts of solar radiation in summer, with a fairly lesser degree in winter when the inclination angle of the sun is lower (between 7° and 28° for most European cities at solar noon). High gains in summer can cause overheating problems if the shading is non-efficient.

Regardless of the quantity, radiation warms the interior after it has been transmitted through the glazing. Such warming is particularly desired in the heating period when it can reduce the heating demand. Free heat gains in large-scale commercial and office buildings covered predominately in glass skin can cause an increase in the cooling demand; special attention thus needs to be paid to the quality of glazing and solar control.

One of the possible approaches to solar collection is to make use of buffer zones which are a type of intermediate sunspaces between the interior and the exterior, usually built in as glazed spaces attached to the southern side of the building. Buffer zones capture direct solar heat gain throughout the day and transfer it to the building's interior by means of natural convection when needed, which usually makes sense in the winter period for warming the interior spaces in the late afternoon and evening, when no direct solar gain is obtainable any more. It is reasonable to use low-e coatings for the glazing of the attached sunspaces in order to prevent heat losses of the long-wave IR radiation (re-emitted from floors, walls and furniture, after they have been heated by the absorbed solar energy) through the glazing back to the exterior. If necessary, sunspaces can in fact supply heat to the building even in spring and autumn but they have to be efficiently shaded in the summer period to avoid overheating (Fig. 2.22). When considering sunspaces, it is important to note that they are seldom occupied and can be therefore exposed to greater temperature variations [11].

Not only the glazing but also the opaque surfaces of the building envelope can collect heat from the sun. When solar radiation strikes the opaque building element, a part of the energy is reflected while another part is absorbed and transformed into heat. As the heat flow transfers progressively towards the internal

Fig. 2.22 Sunspaces can maintain a comfortable indoor temperature in winter nights

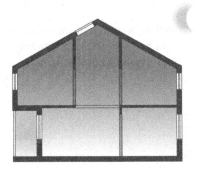

surface by means of conduction, the building element heats up. In addition to the fact that opaque elements do not allow for a direct transmission, they usually have up to three times lower heat transfer coefficient (*U*-value). The latter indicates that the amount of energy passing through the opaque building element is lower than the energy directly transmitted through the glazing.

Another important approach to passive heating strategies is the effect of heat storage in the building materials, which is aimed at retaining the collected heat in order to use it later when required [12]. The latter strategy, which is based on thermal inertia—the capacity of materials to store and release heat, uses the heat stored in the mass of the building materials, i.e., thermal mass, after being struck by solar radiation during the day. The building materials release heat when the surrounding temperature drops. In order to perform effective heat storage, the materials must exhibit higher values of density and thermal capacity. In massive buildings, the structural elements are made of dense materials with high heat storage capacity, e.g., of heavyweight concrete, brick or stone. By absorbing and storing large quantities of heat, such materials can help reduce the temperature swing in the interior spaces during the day. Apart from the above-mentioned traditional materials where the process of sensible thermal energy storage is based on the heat capacity, there exist other materials which can store heat on the basis of phase change (phase-change materials—PCM) or on the basis of chemical processes (thermo chemical materials—TCM). Either concrete or PCM can be used in the interior elements of buildings with lightweight structure, which are not capable of storing heat for a longer time period and therefore react more quickly to the external temperature change. The process of heat storage is beneficial in both, the summer and the winter periods. While it can be used to reduce the interior daytime temperature and to postpone the peak temperature in summer, the advantage of its use in winter is in storing the heat collected during the day and releasing it into space at night.

Another option besides direct heat storage is the use of more complex principles as for instance the Trombe-Michel walls or slabs with an integrated pipe system, where the energy is transferred within or between materials by a heat carrier fluid.

2.6.2 Passive Cooling Strategy

A number of different concepts for passive cooling can be applied to energy-efficient building design including solar control, adequate insulation, the effect of thermal inertia, natural ventilation and natural cooling. Some of these approaches have already been discussed in the previous parts of the text.

2.6.2.1 Heat Storage

Due to *thermal inertia,* the storage of heat and the time lag of heat flow through the building envelope can be used either for heating or for cooling the interior space. The concept is more helpful in climatic regions with significant diurnal variations in temperature. Throughout Europe, there is a relatively large air temperature diurnal swing in the summer period, with even larger swings typical of hot and dry climates. As far as cooling strategy is concerned, the temperature time lag contributes to the reduction and postponement of the interior space peak daytime temperature.

2.6.2.2 Nighttime Cooling

Another important impact of thermal inertia is the fact that the heat stored in building elements can be released to the outside at night, when the external air temperature is usually lower than the internal. This concept is more effective if combined with *nighttime cooling* by means of natural ventilation in order to lower the interior air temperature and precool the building structure for the following day (Fig. 2.23).

2.6.2.3 Cooling by Night Sky Radiation

When the weather is clear, air temperatures at night tend to be very low, which provides a potential for the heat stored in the building materials and roof surfaces or the water from roof ponds to be released to the sky through radiation (Fig. 2.24). This is accompanied by a significant temperature drop of water or other surfaces having radiated heat to the clear night sky. Many different techniques have been tested in order to achieve the effect of night sky radiation cooling. Some of them incorporate using water tanks or pools placed on the roof,

Fig. 2.23 Convective nighttime cooling by means of ventilation

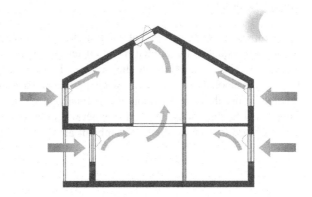

Fig. 2.24 Radiative cooling

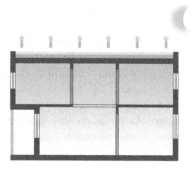

covered by a movable roof during the day. At clear nights, the cover is removed and the exposed water radiates heat to the sky. Such effect can be applied to solar panels where a process opposite to solar heating is used to cool the water temperature, while the electric energy is needed for pumping. Cooling with night sky radiation is less appropriate for humid climates, since humid air is less permeable for long-wave heat radiation.

2.6.2.4 Evaporative Cooling

The concept of evaporative cooling is based on physical process where a substance absorbs the energy known as latent heat of vapourization to transform its state from liquid to gas. In the case of evaporative cooling, the heat supplied by hot air causes transformation of the water from fountains, pools and ponds, located next to the building or on its roof, to vapour. Consequently, a drop in air temperature and an increase in air humidity appear. The long-wave IR radiation process described above could also be applied in the building interior where it is necessary to control the level of humidity. However, the evaporative cooling effect cannot be applied in regions with a humid climate.

2.6.2.5 Ground Cooling

The temperature of the ground at a depth of approximately 3–4 m, which has been found to be equal to the annual mean air temperature for the location [11], might vary by ±2 °C depending on the season. The air used for ventilation of the building is cooled by a passage through the underground duct. Although it is possible to use this effect without any mechanical systems, ground cooling is more often used within active systems, where ventilation units are combined with ground pipes to precool the fresh ambient air (Fig. 2.25).

Fig. 2.25 Ground cooling

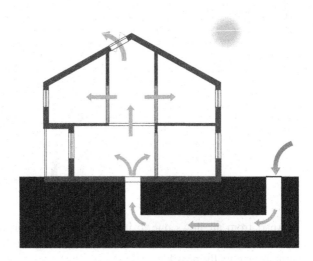

2.6.3 Natural Ventilation

Natural ventilation is primarily used to supply fresh air to the building. On average, a human being needs a minimum of 8 l of fresh air per second. An adequate fresh air provision depends on the number of persons in the room and their activity. To assure the physiological needs and to maintain the internal air quality, fresh air can be supplied either by natural or mechanical ventilation. With natural ventilation, it is necessary to control the outdoor air temperature to prevent larger infiltration heat losses or gains which could appear in case of a large difference between the interior and exterior temperature. To avoid substantial ventilation heat losses in summer and winter, the total fresh air supply is usually provided by mechanical ventilation.

As already mentioned in the previous subsection, natural ventilation is also beneficial in the summer time for the reason of ensuring thermal comfort. Significantly larger ventilation rates (80–100 l per person per second) are necessary to cool the building, which, however, depends on several factors like the external air temperature, exposure of transparent surfaces to solar radiation, internal gains, etc.

In general, natural ventilation is driven by either wind or thermal forces [11]. While the wind-driven ventilation is induced by the pressure differences, thermal circulation caused by a temperature difference between the outside and inside air induces a natural flow through the building. The direction of the flow depends on the temperatures. The air mass inside the building having a higher temperature than that of the outside air has lower density than the outside air mass and tends to move upwards, while the cooler air from the outside flows into the lower areas of the building. This type of ventilation, called "stack ventilation" or "chimney effect", is most efficient if operable windows are installed at the bottom and the top of the building (Fig. 2.26).

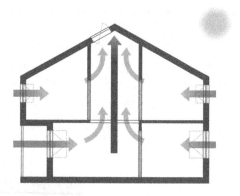

Apart from air flows due to infiltration of fresh air, the appearance of the air
mass flow between different areas of the building due to their different air temperature is equally possible.

On the other hand, the wind-driven natural ventilation can be provided through
single-sided or cross-ventilation. Induced by a difference in the wind pressure that
usually arises between the windward and the leeward sides of a building, it is most
effective if the air mass is driven through the building as cross-ventilation. Single-
sided ventilation is beneficial for the provision of fresh air as well, but achieves a
rather lower air exchange than cross-ventilation (Fig. 2.27).

Different types of ventilation can be combined to achieve an even better effect.
In regions with constant winds, the positioning of buildings has to be carefully
considered. Besides the provision of fresh air, wind forces contribute to decreasing
the surface temperature due to convection, which can be beneficial for cooling the
building mass. If there is a need to reduce the wind impact, certain barriers made
of vegetation, walls or fences should be designed at exposed areas. These can also
be used to divert the wind direction and achieve better ventilation in cases where
the building openings cannot be positioned on the exposed façades (Fig. 2.28).

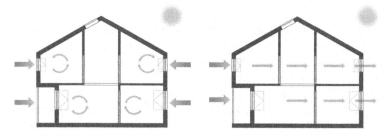

Fig. 2.27 Single-sided and cross-ventilation

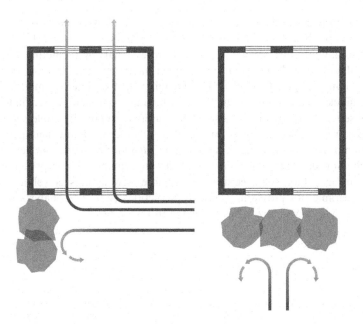

Fig. 2.28 Vegetation barriers used to the divert the wind direction where needed

2.6.4 Daylighting

Daylighting is a design strategy employing the available daytime visible light to illuminate the interior space. Daylight has been used for centuries as the primary source of light in interiors and has been an implicit part of architecture for as long as buildings have existed [26]. Nowadays, when we spend 90 % of our time indoors, adequate daylight has become even more important. Several studies have proved the value of daylight providing visual comfort benefits that are essential for improved productivity and satisfaction of the buildings' occupants. Properly applied daylighting prevents inconvenient glare effects by not allowing the sun to directly enter a space and ensures good natural illumination across the entire internal space [1]. Besides ensuring an adequate visual comfort, daylight also reduces the need for artificial electrical lighting during daytime hours.

Daylight can reach the interior space through the glazed openings in the roof or through the façade windows. The roof windows are usually limited to the top floor of a building, while the façade windows can be applied to multiple floors of a building, satisfying the requirements of correct exposure and orientation [1]. An integrative approach combining both, the roof and façade windows properly distributed in the window envelope, is advised for the achievement of optimal lighting comfort. Additionally, the openings should be equipped with shading devices for situations when daylight is too bright or when a direct sun causes glare problems.

For the purposes of daylight building design, several tools can be of assistance. One of such freely accessible tools is VELUX Daylight Visualizer 2. It is intended to analyse daylight in buildings and to aid professionals by predicting and documenting daylight levels on the one hand and visualizing a space prior to realization of the building design on the other. The daylight visualizer intuitive modelling tool permits quick generation of 3D models in which the roof and façade windows are freely inserted. The programme also enables users to import 3D models generated by CAD programmes in order to facilitate a good workflow and permit the evaluation of a wide range of building designs, in addition to offering flexibility in the evaluation process. Other features of the programme include predefined settings, a surface editor, site specifications, flexible view settings as well as multiple daylight parametrics providing accurate predictions [26]. The analysis made by VELUX Daylight Visualizer 2 is presented in Fig. 2.29.

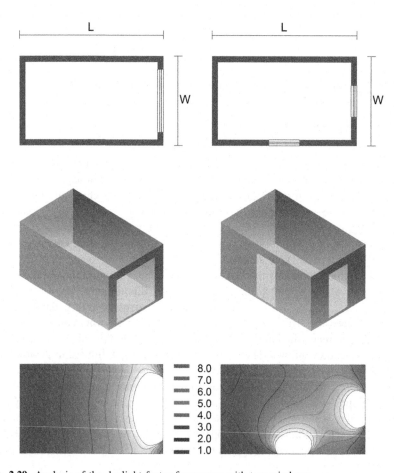

Fig. 2.29 Analysis of the daylight factor for a room with two windows

The above figure demonstrates the influence of the window openings (bright spots) exerted on the overall daylight conditions in the room.

The daylight factor, as one of the indicators the programme uses to assess daylight quality, describes the ratio between the amount of daylight available in the interior (at the height of the work plane) and the amount of non-obstructed daylight available outside under standard CIE overcast sky conditions. It is expressed as a percentage [15].

The daylight factor is common and easy to use measure for the available amount of daylight in a room. It can be measured for a specific point or expressed as an average. The latter is the arithmetic mean of the sum of point measurements taken at a height of 0.85 m in a grid covering the entire floor area of the room. The higher the DF, the more daylight is available in the room. An average DF below 2 % generally makes a room look dull and electrical lighting is likely to be frequently used. Rooms with an average DF of 2 % or more can be considered day lit, but electrical lighting may still be needed to perform visual tasks. A room will appear strongly day lit when the average DF is above 5 %, in which case electrical lighting will most likely not be used during daytime [4].

2.7 Active Technical Systems

Besides passive building components presented in previous subsections, active technical systems are necessary for a complex integrated building operation. Active technical systems refer to heating, ventilation and air conditioning (HVAC), domestic hot water supply, artificial lighting and renewable energy systems. They all need to consume electrical power for their performance. With the exception of lighting, these processes can be classified as mechanical systems. A way to improve the overall efficiency of contemporary mechanical systems is to incorporate strategies that use surrounding natural conditions like outdoor air, solar radiation, the ground or groundwater.

An example of such strategy can be demonstrated by the energy recovery ventilation system. In energy-efficient houses, it is necessary to combine natural and mechanical ventilation to ensure suitable indoor air quality. In order to minimize ventilation heat losses, a heat recovery system (HRV) can help make mechanical ventilation more effective by reclaiming energy from exhaust airflows. The system uses the conditioned exhaust air to precondition the incoming fresh air that needs to be heated or cooled. Both, the exhaust and fresh air pass through a heat exchanger where the incoming air is preheated or cooled by the energy of the exhaust air, thereby reducing the amount of conditioning needed. In a typical system configuration, air is supplied to the living room, dining room and bedrooms while it is removed from the kitchen, bathroom and toilets.

Electricity consumption can be minimized by the use of highly efficient systems, in addition to energy-efficient household appliances, light bulbs and luminaire. The use of photovoltaic panels (PV) intended for the production of electrical energy are a further improvement of the overall energy balance.

With efficient renewable energy systems, such as heat recovery systems, heat pumps, solar panels, photovoltaic panels, etc., providing ventilation, space heating, domestic water heating and even electricity, the building uses less or even no fossil fuel for its operation, which leads to lower environmental burdening.

The main goal of energy-efficient house design is to reduce the overall energy use in the building primarily through designing the optimal building shape, orientation and building components as well as by optimizing passive strategies. The next stage is to adapt the specific energy-efficient technical systems to the existing building design. Since none of these strategies will result in maximum efficiency without the cooperation of the buildings' owners, operators and occupants, it is important to educate all the participants on the proper use of energy-efficient buildings and their technology. With respect to all these parameters, energy-efficient building design is understood as an extremely complex process which demands an accurate approach in planning and selecting each of the individual parameters.

References

1. AIA (2009a) Daylighting. Available via: http://wiki.aia.org/Wiki%20Pages/Daylighting.aspx
2. AIA (2009b) Sun shading. Available via: http://wiki.aia.org/Wiki%20Pages/Sun%20Shading.aspx
3. BEMBook (2012) Solar shading, In: http://www.bembook.ibpsa.us/index.php?title=Solar_ Shading#Exterior_shading_device
4. CIBSE (2002) Code for lighting. Chartered Institution of Building Services Engineers, Oxford
5. DIN V 4108–6 (2003–06) Thermal protection and energy economy in buildings—Part 6: Calculation of Annual Heat and Annual Energy Use
6. DIN V 18599–2 (2007–02) Energy efficiency of buildings—calculation of the net, final and primary energy demand for heating, Cooling, Ventilation, Domestic hot water and lighting— Part 2: Net energy demand for heating and cooling of building zones
7. Dixon JM (2008) Heating, cooling, and lighting as form-givers in architecture, In: Lechner N (2008) Heating, Cooling, Lighting, Sustainable design methods for architects, 3rd edn, Wiley
8. EnEV Energieeinsparverordnung für Gebäude (2009)
9. European Commission (2009) DG energy and transport, directorate D: Low energy buildings in Europe: current state of play, definitions and best practice
10. Feist W (2012) Passive house planning package (PHPP 1998–2012), Energy balance and passive house design tool for quality approved passive houses and EnerPHit retrofits. Passive House Insitute, Darmstadt
11. Ford B, Schiano-Phan R, Zhongcheng D (2007) The Passivhaus standard in European warm climates, Design guidelines for comfortable Low-energy homes—Part 3: Comfort, Climate and passive strategies, Passive-On project report
12. Goulding JR, Lewis JO, Steemers T (1992) Energy conscious design, A primer for architects, Written by: Architecture et climat, Centre de researches en architecture, Université Catholique de Louvain, Belgium, Produced and coordinated by: The energy research group, School of Architecture, University College Dublin, Batsford Ltd, London

13. Hachem C, Athientitis A, Fazio P (2011) Parametric investigation of geometric form effects on solar potential of housing units. Sol Energy 85:1864–1877
14. Hammond GP, Jones CI (2008) Embodied energy and carbon in construction materials. Proceedings of the ICE—Energy 161(2):87–98
15. Hopkins RG (1963) Architectural physics: lighting. Her Majesty's Stationary Office, London
16. Jelle BP, Gustavsen A, Baetens R (2010) The high performance thermal building insulation materials and solutions of tomorrow, In: Proceedings of the thermal performance of the exterior envelopes of whole buildings XI international conference (Buildings XI), Clearwater Beach, Florida, U.S.A., 5–9 Dec
17. Jelle BP (2011) Traditional, State-of-the-Art and future thermal building insulation materials and solutions—properties, requirements and possibilities, Energ Build 43:2549–2563
18. Kottek M, Grieser J, Beck C, Rudolf B, Rubel F (2006) World map of the Köppen–Geiger climate classification updated. Meteorol Z 15:259–263. doi:10.1127/0941-2948/2006/0130
19. Laustsen J (2008) Energy efficiency requirements in building codes, Energy efficiency policies for new buildings. IEA Information paper, In Support of the G8 plan of action, International Energy Agency © OECD/IEA
20. OIB Richtlinien 6 (2011) Energieeinsparung und Wärmeschutz, Österreichisches Institut für Bautechnik
21. Papadopoulos AM (2005) State of the art in thermal insulation materials and aims for future developments. Energ Build 37:77–86
22. Peel MC, Finlayson BL, McMahon TA (2007) Updated world map of the Köppen–Geiger climate classification. Hydrol Earth Syst Sci 11:1633–1644
23. Rules on the methodology of construction and issuance of building energy certificates (2009) Official Gazette RS, 77/2009
24. Szokolay SV (2008) Introduction to architectural science: The basis of sustainable design. Elsevier, Oxford
25. Vitruvius (1960) The ten books on architecture (Bks. I–X)
26. VELUX Group: Andersen PA, Duer K, Foldbjerg P, Roy N, Andersen K, Asmussen TF, Philipson BH (2010) Daylight, Energy
27. http://www.timeanddate.com/
28. http://windows.lbl.gov/daylighting/designguide/section5.pdf

Chapter 3
Structural Systems of Timber Buildings

Abstract The current chapter presents a set of main aspects of timber building with a focus on timber-frame constructions. A brief description of timber's material characteristics in Sect. 3.1 aims at getting acquainted with potential advantages and disadvantages of planning and designing timber buildings, in comparison with using other structural materials, such as concrete or masonry. Section 3.2 discusses predominantly used structural systems of timber construction along with the most important structural and technological characteristics and possibilities. Section 3.3 describes computational models and methods, with their limitations and application in practice. The influence of the sheathing material and the openings studied in our previous numerical and experimental research is additionally shortly presented in order to provide a better insight into a rather problematic area of applying the glazing to timber buildings, which is the main contents part of Chap. 4. Multi-storey prefabricated timber building is one of the increasing opportunities for the public, commercial and residential sectors in the future. Stability problems appearing due to heavy horizontal actions along with possible strengthening solutions already applied in practice are the topic of Sect. 3.4.

3.1 Timber as a Building Material

Timber is a live organisms' product and thus a natural material exposed to parasites and bacteria. Alternate exposure to humidity makes timber unsustainable while its organic structure accounts for its *inhomogeneity,* which is a rather negative construction-related feature. Another specific area is timber's *fire resistance,* a highly specific problem to which a more detailed approach follows further in this chapter. The three characteristics mentioned above are said to be the main drawbacks of using timber in construction. However, the listed disadvantages can

V. Žegarac Leskovar and M. Premrov, *Energy-Efficient Timber-Glass Houses,*
Green Energy and Technology, DOI: 10.1007/978-1-4471-5511-9_3,
© Springer-Verlag London 2013

be partially or even fully overcome with appropriate use of timber, which will be further discussed in Sects. 3.1.1, 3.1.2 and 3.1.3.

Timber has, on the other hand, excellent construction features. Its compressive strength is almost equal to that of concrete but its tensile *strength* is significantly higher. The most important advantage over concrete is its much *lower weight*. Moreover, if the weight of both materials is equivalent, timber satisfies almost the same construction requirements as steel. Nevertheless, on account of its relatively low value of the modulus of elasticity, which is approximately three times lower than that of concrete and twenty times lower than that of steel, timber is not suitable for structures with extreme spans or heights although it has become an ever more frequent material used in multi-storey prefabricated construction, which is a topic treated in Sect. 3.4.

Timber is, in addition, a non-demanding material for prefabrication due to its organic structure and low density. It is also an ideal construction material from the viewpoint of *energy efficiency* since CO_2 emissions in production of a timber element tend to be approximately two times lower than that those present in manufacturing an equivalent brick element, three times lower than in the case of a concrete element and six times lower than CO_2 emissions in steel element production. The reason lies in photosynthesis enabling a growing tree to store CO_2 which is then released only in the burning process of the timber mass. Since the above-described characteristics of timber frequently reoccur in Chap. 4, the following subsections aim at their more specific presentation.

3.1.1 Inhomogeneity of Timber

Timber is an organic substance, inhomogeneous in the organic, anatomical and physical sense. Most of its physical and mechanical properties differ depending on the grain direction, which is seen in the pronounced anisotropy. Timber's properties are generally best in the direction parallel to the grain, with their intensity decreasing proportionally to the deviation from the longitudinal axis, reaching bottom qualities perpendicularly to the grain. These are typical features distinguishing timber from other construction materials.

A detailed analysis of wood structure needs to be preceded by a definition of a set of terms to be used in our further discussion. In contrast to some tropical and subtropical trees (palm tree, bamboo), European trees' growth is characterized by a simultaneous increase in the tree height and width, with the height growth being typical of the juvenile phase followed by the diameter growth in the full vigour phase. The latter results in cylinder-shaped growth layers called annual rings. Clearly visible boundaries between the layers are called annual ring boundaries. They appear as concentric circles arranged around the stem core called the pith in the cross section and look like axial, almost parallel lines (Fig. 3.1) in the radial section. Annual ring boundaries typically have a more distinctive definition in conifers than in deciduous trees. The width between annual ring boundaries is referred to as annual ring width.

As seen in the cross section, annual ring width becomes larger every year which is seen in progressive distribution of the annual rings ranging from the narrowest ones around the pith where timber is the oldest and most compressed to the widest annual rings on the bark side where timber is the youngest. With annual ring width becoming larger the levels of timber density, strength and elasticity decrease proportionally to the distance from the pith. Timber whose annual ring width exceeds 5 mm tends to be rather soft and is normally not used for load-bearing elements. The figure also clearly shows larger distances between annual ring boundaries in the tangential section as compared to the radial section. Timber strength is thus slightly higher in the radial than in the tangential direction.

Timber is a natural material and the pace of growth in conifers differs from that in deciduous trees. It is generally true that conifers grow faster which is the reason for deciduous trees to have narrower annual ring widths. As a consequence, the density (as well as the strength) of deciduous trees is on average higher (than that of conifers). In addition, there is a difference in the seasonal growth pace of conifers and deciduous trees. Conifers grow faster in spring than in autumn, while the opposite is true of deciduous trees (Fig. 3.2).

Observing an individual annual ring (Fig. 3.2) leads to an additional conclusion claiming that timber behaves as inhomogeneous material. Spring (early) timber of conifers tends to be more porous than autumn (late) timber and the strength of spring timber of conifers is thus lower. The reverse is true of deciduous trees whose spring timber is less porous and has higher strength than autumn timber. The difference between spring and autumn growth is slightly more evident in conifers.

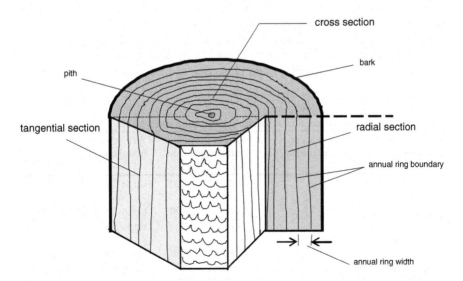

Fig. 3.1 Cross section of the wood element

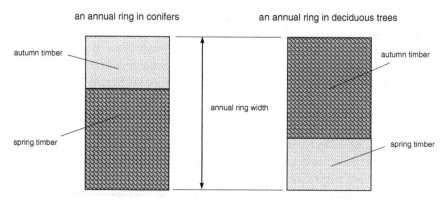

Fig. 3.2 Spring and autumn growth in conifers and deciduous trees

Timber-stem-structure-related descriptions above prove the fact that timber is a rather inhomogeneous material. Its inhomogeneity is seen not only in the stem structure following individual annual ring boundaries ranging from the pith and out but also inside the annual ring (annual ring width) itself. Inhomogeneity in the cross section of the stem is seen in a decrease in density from the pith to the bark, while it depends on the type of tree (conifers or deciduous trees) within a single annual ring width and differs in spring and autumn growth periods.

3.1.2 Durability of Timber

Durability is defined as a period during which most natural properties (anatomical structure, colour, strength, etc.) of timber remain unaltered. Durability of timber is subject to changes since it ranges from a few months to a hundred years and even more (e.g. timber piles in Venice). Durability depends primarily on weather conditions (changes in humidity) and protection against humidity as well as on

Table 3.1 Durability of certain timber species (by Campredon) (in years)

Timber species/ environment	In permanent ground contact	Without ground contact open under roof		Well protected from humidity	Soaked in fresh water
Oak, chestnut, elm, hornbeam,	8–12	60–120	>200	Up to 500	>500
Ash, birch, maple,	4–6	20–60	>100	Up to 500	50–100
Beech, poplar, willow, linden	<4	<30	>50	Up to 500	<50
Pine	8–12	40–80	>150	Up to 500	>500
Fir, spruce fir	<4	30–50	>50	Up to 500	<50

timber species. Table 3.1 lists Campredon's average durability values for raw timber of different species, in dependence on climate conditions. Factors of timber protection to increase its durability are not taken into account.

As seen in table, the environment where timber once cut is stored plays a vital role. Changes in humidity which mostly affect elements in permanent contact with the soil cause significant decrease in timber durability. On the contrary, timber kept in places well protected from humidity will retain its properties.

Data referring to timber permanently soaked in fresh water seem to be of particular interest. They point to durability of certain species (oak, chestnut, elm, hornbeam, pine) which is practically limitless while that of other species (beech, poplar, willow, linden, spruce fir, and fir) does not exceed 50 years. It follows that the most decisive factor in reducing timber's durability is not permanent water content but the *change in humidity*. Therefore, if we want timber to have a long life span, we need to protect it from humidity changes and not from permanent water content.

Durability also strongly depends on timber species. Based on the above data, timber species could be classified into three different durability groups:

High-durability species	Good-durability species	Low-durability species
Oak	Spruce fir	Beech
Elm	Fir	Maple
Hornbeam	Pine	Birch
	Ash	Linden
	Chestnut	Poplar
		Willow

It would be difficult to set a standard defining durability of conifers on the one hand and deciduous trees on the other. Nevertheless, there is rule of not using low-durability species of timber in construction but only for manufacturing furniture and similar elements which are well protected from humidity changes throughout their exploitation period. In contrast, we mostly use high- and good-durability timber species solely for the construction purposes.

All in all, a matter of utmost importance is proper drying of timber to be used in construction when it needs to have the expected humidity level. The latter must be lower than the admissible level of humidity defined by regulations (based on European regulations to be applied in particular cases of the use of timber), which deprives pests from one of their basic life requirements.

3.1.3 Fire Resistance of Timber Structures

Lack of fire resistance is said to be one of the main deficiencies of timber structures, which is a reproach based on insufficient knowledge of the behaviour typical of timber elements under fire load. Appropriate design and planning of timber elements can almost completely overcome these alleged disadvantages.

Fire characteristics of construction elements can be defined by the following parameters:

- Combustibility
- Smoke content level
- Toxicity
- Material decay due to internal tensile stress
- Ability to change phases—from solids to liquids or gases.

The listed parameters are not all equally important for timber as a construction material. Factors of toxicity, fusibility and decay classify timber ahead of other construction materials, while the factors of combustibility, expanding flame speed and smoke content level undoubtedly place timber in a subordinate position. Timber is organic material characterized above all by inflammability and combustibility. It consists mainly of cellulose and lignin, two organic substances containing a high percentage of carbon, which explains why insisting on the definition of the onset of timber burning at increased outside temperature is wrong since the concept of combustibility cannot be treated separately from that of timber oxidation. The presence of the latter phenomenon at any temperature accounts for the change of timber colour at normal day temperatures. Timber behaviour at temperatures below 275 °C has not been researched enough but timber changes are obvious. Timber loses its weight and undergoes a pronounced colour change, which can also be merely a consequence of certain chemical reactions.

Timber is a combustible material causing a high smoke content level in the building on fire. Combustibility is the reason why people might feel reluctant to use timber in construction, especially in Central Europe but less in Scandinavian countries and overseas—in Canada and the USA. Nevertheless, as far as *toxicity* and *decay* are concerned, timber has considerable advantages over other materials.

During the process of burning, it forms a charred layer on the surface, a layer having a self-protective function (Fig. 3.3). The usual layer thickness of 5–10 mm is generally very small in comparison with dimensions of the cross section of a substantial size. Mechanical properties of timber exposed to high temperatures can thus remain almost unaltered for a longer period of time than those of other materials but the problem of smoke causing suffocation remains unsolved. Furthermore, renovation of such timber elements is rather simple. A charred layer can be easily removed and all the steel connections need to be replaced, which was evident in the case of renovating the small thermal swimming pool at Zreče Thermal Spa, located at the foot of the Pohorje Mountain, Slovenia (Fig. 3.4).

As inferred from the above statements, timber has a higher fire load capacity than concrete or steel, i.e., it reaches its yielding point much later. Timber structures usually remain non-deformed in the fire aftermath, which cannot be said of steel structures. The proof is seen in numerous cases of fire damage observed on timber and steel structures, in addition to instances of timber and steel elements being

Fig. 3.3 Charring of the glued-laminated beams after a fire incident in Hotel Dobrava (Zreče Spa)

Fig. 3.4 Charring of the glued-laminated beams and bending of the steel beams

located close to each other and thus exposed to the same fire and temperature loads. It would be therefore incorrect to define timber structures as non-resistant to fire.

A clear example in support of the previous claim was seen in the consequences of a large fire at two of the pools in the Dobrava Hotel at Zreče Thermal Spa, on 8 April 2001, where the primary timber elements suffered no extensive damage. Glued-laminated beams above the pools were completely charred (Fig. 3.3), which was an unpleasant sight suggesting the roof structure was unusable. Nevertheless, the removal of the upper charred layer showed that elements suffered only 5 mm outer surface damage, which could have been observed on the paned parts of the primary glulam beams (Fig. 3.3). The outer surface charred layer acts as protection of the inner parts that remain unburnt.

Fig. 3.5 Renovated primary
glued-laminated beams in the
smaller thermal pool

A detailed examination showed no other damage except for the charring of the glued-laminated beams, not even the increase in bending. It can be subsequently concluded that there was no major decrease in the load-bearing capacity of the beams. On the other hand, steel components sustained certain deformation at individual connections, which is an outcome that could affect the static system. Unfortunately, steel beams underwent heavy deformation whose occurrence in such mixed systems is not unusual (Fig. 3.4).

The load-bearing capacity of the primary glued-laminated beams did not significantly decrease and made the renovation quick and easy since it only encompassed the removal of the charred layer (Fig. 3.5). Replacing the beams would have been unnecessary from the economical point of view, however, all steel connections definitely needed replacement (Fig. 3.6).

Fig. 3.6 Renovated steel
connection

Secondary roof beams along with their timber panelling, in addition to roofing and insulation were completely replaced or renovated. The process of renovation included problematic steel connections, where all damaged steel parts were replaced with new ones (Fig. 3.6).

The renovation of the fire-damaged glued-laminated beams and the damage sustained by metal roof beams presented above serve as a proof that timber structures cannot be classified as non-resistant to fire. On the contrary, proper design and planning (cross sections of adequate size) ensure a sufficient fire-resistant level. Methods for planning and designing fire-resistant structures are defined in »Eurocode 5, Part 1–2: General rules—Structural fire design« which prescribe three alternative computation methods to ensure fire resistance of timber structures.

3.1.4 Sustainability of Timber

Being a natural raw material requiring minimal energy input into the process of becoming construction material, timber represents one of the best choices for energy-efficient construction, since it also functions as a material with good thermal transmittance properties if compared to other construction materials. Moreover, timber has good mechanical properties and ensures a comfortable indoor climate in addition to playing an important role in the reduction in CO_2 emissions. Trees absorb CO_2 while growing (estimated CO_2 absorption of conifers is approximately 900 kg per 1 m^3 with that of deciduous trees being 1,000 kg per 1 m^3), which makes timber carbon neutral; a building made of an adequate mass of timber can thus have even a negative carbon footprint.

Table 3.2 shows the grey energy consumption, also called LCA or a "cradle-to-grave analysis", of a selection of most frequently used building materials and their end product elements. As the density of the materials varies, the values of the energy consumption per kg and m^3 are given separately.

Table 3.2 Grey energy consumption for various building materials

Building material	Grey energy (MJ/kg)	Grey energy (MJ/m^3)
Aluminium	191–227	517,185–611,224
Aluminium recycled	8.1–42.9	24,397
Steel	31.3–74.8	245,757–613,535
Cement	5.2–7.8	12,005–12,594
Brick	2.5–7.2	5,310–14,885
Glass	15.9	40,039
Insulation–polystyrene	117	1,401
Timber	0.3–1.6	165–638
Wood-based boards (MDF, OSB)	8.0–11.9	5,720–5,694

It is evident from the data above that the grey energy consumption in producing 1 kg of the timber element is the lowest of all, having a value nearly 5 times lower than that of brick, 6 times lower than in the case of cement and approximately 50 times lower in comparison with steel.

It is also interesting to compare the data for CO_2 emissions assessed in the production of 1 m^2 of timber wall elements on the one hand and the same size of brick wall elements on the other, where the same type of insulation is inserted. Manufacturing 50 m^2 of timber wall elements will emit around 1.5 tonnes of CO_2, a quantity that roughly amounts to 5 tonnes in the case of brick wall elements. It is therefore clear that using timber in the construction of residential, commercial and public buildings leads to substantial reductions in CO_2 emissions.

3.1.5 Timber Strength

One of the biggest advantages of timber is definitely its relatively high strength in respect to its rather low density. A comparison of the moduli of elasticity and density shows that a ratio of timber is twice as favourable as that of concrete, while a comparison of their compressive strength proves an even better ratio in favour of timber.

Example A comparison of timber of the strength class C30 according to Ref. [35] and concrete of the strength class C30/37 according to Ref. [34]. Basic material characteristics are given in Table 3.3.

Based on the calculated results, the compressive strength to density ratio shows a 4.16 times higher value in timber, while the modulus of elasticity to density ratio proves to be 2.07 times higher in timber. As far as the modulus of elasticity is concerned, it needs to be pointed out that it is 3 times lower than that of concrete, which certainly assigns a subordinate position to timber when it comes to structures with extreme spans.

Since Chap. 4 lays a stronger focus on selected timber strength characteristics as, only basic property details of the tensile, compressive and bending strength of timber in addition to the modulus of elasticity and the shear modulus follow below.

Table 3.3 Material characteristics and calculated results

	Density ρ (kg/m^3)	Compressive strength f_c (N/mm^2)	Ratio f_c/ρ	Modulus of elasticity E (N/mm^2)	Ratio E/ρ
Timber C30	460.00	23.00	0.050	12,000	26.09
Concrete C30/37	2,500	30.00	0.012	31,500	12.60
Timber/concrete (%)	18.40	76.67	416.67	38.10	207.06

3.1.5.1 Compressive Strength

The compressive stress parallel to the grain appears if the compression force acts lengthwise (Fig. 3.7). The stress causing timber element to break is called the compressive strength parallel to the grain.

Basic elastic and plastic properties of timber under stress are shown by the σ-ε diagram for pine (Fig. 3.8).

The diagram shows relatively ductile behaviour of timber in compression. Until reaching the point of proportion (A), at approximately 50 % of the compressive strength, timber demonstrates fully elastic behaviour. Higher compressive stress leads to more extensive deformation, which is seen in increasingly plastic behaviour of the material until the point of failure occurring at specific deformation of approximately ($\varepsilon = 7$ ‰). Furthermore, the compressive strength of timber under long-term load ($f_{c,0,t=\infty}$) is considerably lower than its strength under instantaneous load ($f_{c,0,t=0}$). The ($f_{c,0,t=\infty}$)/($f_{c,0,t=0}$) ratio is approximately 55–65 %, which is contained in the coefficient k_{mod} according to Eurocode 5.

In the case of timber exposed to dynamic load, it would be sensible to define its dynamic compressive strength. The latter is relatively high (nearly 90 %) as compared to the static compressive strength owing to mainly ductile material behaviour, which confirms the suitability of using timber under dynamic compressive loads.

Fig. 3.7 Compressive stress parallel to the grain ($\alpha = 0$)

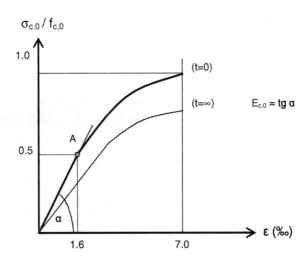

Fig. 3.8 σ-ε diagram of timber in compression parallel to the grain

3.1.5.2 Tensile Strength

The tensile strength parallel to the grain is defined as resistance of the material to the tensile stress acting lengthwise (Fig. 3.9).

The tensile strength is transferred along the grain and thus strongly affected by irregularities in the timber structure. Knots are certainly a feature highly detrimental to the tensile strength. The tensile stress perpendicular to the grain highly concentrated in the knot and around it causes tensile failure even in the case of low-load exposure (Fig. 3.10). The failure occurring without any previous signs is generally extremely fast since timber is not ductile in its tensile zone.

Knots in timber will significantly reduce its tensile strength but will cause a substantially lower reduction of its compressive strength. Every knot represents a local change in the inclination of the grain towards the element's axis simultaneously with the reduction in the load-bearing area of the section.

The above facts can consequently cause substantial deviation in defining the tensile strength of timber. The behaviour of timber in tension can be best shown by the σ-ε diagram. Figure 3.11 presents a diagram for the pine tree.

The diagram exhibits rather non-ductile behaviour of timber in tension as compared to compression. The computed value of the point of proportion is set relatively high (at 90–95 % of the tensile strength); nevertheless, a slight distortion of the stress–strain diagram appears already at 50 % of the tensile strength. Irreversible distortion at higher values of the tensile stress is in fact small since the stress curve only slightly deviates from the proportional straight line. Tensile failure occurs at the approximately same strain as compressive failure.

In the case of timber exposed to dynamic load, it would be sensible to define its dynamic tensile strength. On account of rather non-ductile behaviour of timber in tension, the value of its tensile strength is relatively low as compared to that of the

Fig. 3.9 Tension parallel to the grain

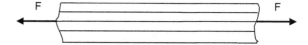

Fig. 3.10 Tensile failure due to a knot

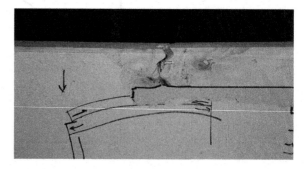

Fig. 3.11 σ-ε diagram of timber in tension parallel to the grain

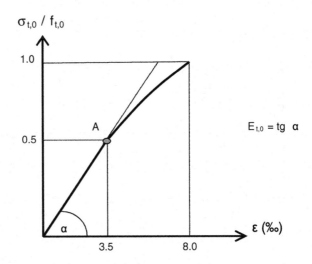

static tensile strength (only about 40–60 %), which is a lot less than in the case of the compressive strength.

3.1.5.3 Bending Strength

The modulus of elasticity of timber up to the point of proportion is nearly the same for timber in tension or compression, which is why the previously mentioned differences between the tensile ($f_{t,0}$) and compressive ($f_{c,0}$) strength of timber, i.e., between their σ-ε diagrams, do in fact define the bending strength of timber (f_m). Following the Bernoulli–Euler hypothesis applicable to homogeneous and isotropic elements, in addition to assumed elastic behaviour of the material (Hook's law), there exists a linear disposition of the compressive and tensile stresses in the cross section, which we will name *Phase I* (Fig. 3.12a). Accordingly, failure of the timber element under bending load should occur when the maximum stresses reach the tensile or compressive strength with the decisive value being that of the lower one. Yet, the fact that the above disposition of stresses for homogeneous and isotropic elements, such as timber, exists only up to the point of proportion needs to be taken into consideration.

With higher loads acting upon the element, the compressive stress tends to be less proportional to specific deformation and its disposition in the cross section is thus no longer linear (Fig. 3.12b, c). Likewise, the modulus of elasticity of timber in tension and that of timber in compression are no longer the same. The yielding first occurs in the area of the lower point of proportion. As seen in the previous chapters, the behaviour of timber in the compressive zone is of higher ductility than that of timber in the tensile zone, which makes the point of proportion in the compressive zone nearly always lower than the limit point of proportion in the tensile zone. Figure 3.12b thus shows the so-called intermediate phase

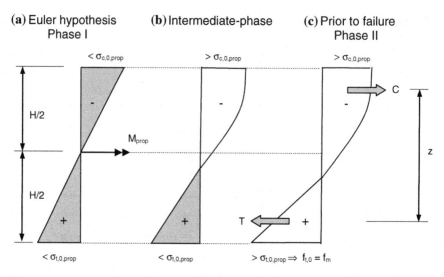

Fig. 3.12 Stress state of the timber element until bending failure

demonstrating nonlinear behaviour of timber in the compressive zone, whereas compression in the tensile zone displays a linear course due to a higher point of proportion. With further load increase, the neutral axis shifts to the tensile zone, while the yielding of timber occurs in the tensile zone as well—*Phase II* (Fig. 3.12c). In order to achieve balance, the resultants of the compressive and tensile strength need to be the same $(T = C)$. The moment of internal forces $(M = T \cdot z)$ needs to be the same as that of external forces.

Further increase in the load causes progressive shifting of the neutral axis towards the tensile edge until the maximal stress reaches the compressive or tensile strength of the material, which leads to the failure of the element. The resultant force of the stress zone in the failure area equals zero and does consequently not ensure the balance. The stress causing failure in the tensile or compressive zone is called the bending strength of the material.

The bending stress acting on the radial or tangent plane defined by the annual ring boundaries direction of the element is referred to as radial or tangent bending, respectively (Fig. 3.13). In both cases, the normal stress (σ_x) is parallel to the grain.

Radial bending occurs when the load plane is radial to annual ring boundaries, while with tangent bending, the load plane is tangential to annual ring boundaries. The radial bending strength is higher than the tangential $(f_{m,\mathrm{rad}} > f_{m,\mathrm{tang}})$ as timber is more compressed in its radial direction. It is thus recommended, within a scope of possibility, to place the element radially to the load direction.

Fig. 3.13 Radial and tangent
bending

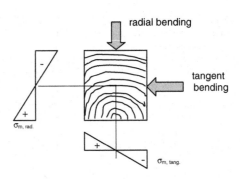

3.1.6 Modulus of Elasticity

Like timber strength, the modulus of elasticity depends mainly on timber species, growth related irregularities, density, porousness, humidity and grain direction. Timber with a regular pattern of annual ring boundaries and grain has a higher elasticity value than timber with grain deviation or uneven annual ring boundaries width. Increased humidity exerts a negative effect on timber's elasticity, as is also the case with timber strength. Furthermore, due to rheological phenomena, elasticity decreases with the ageing of timber.

The modulus of elasticity is a mechanical property describing timber's elasticity which is also referred to as *Young's modulus*. The modulus of elasticity is mathematically defined as the ratio of the increment of the stress ($d\sigma$) to the differential of specific deformation ($d\sigma$):

$$E = \frac{d\sigma}{d\varepsilon} \qquad (3.1)$$

It follows from the above equation that a linear ratio of the stress to specific deformation results in a constant modulus of elasticity $E = \sigma/\sigma$, which is referred to as the elasticity area of the material's behaviour. Upon removing the load actions, the element under ideally applied elastic stress returns into its original position and elastic deformation is hence also called reverse deformation, since energy itself is reversible. Increasing the stress reduces the linearity of the ratio σ-ε, which means that the modulus of elasticity expressed by the Eq. (3.1) is no longer a constant as its value slowly decreases. The element under such stress gradually loses its stiffness and undergoes the process of yielding until the point of failure. Deformation is not reversible upon the removal of the load actions; it is irreversible within the area of yielding. Figure 3.14 is a schematic diagram showing elasto-plastic material behaviour.

Timber's organic structure accounts for its anisotropy and inhomogeneity, which results in the modulus of elasticity being dependent on:

- The type of stress (compression, tension, bending)
- The stress-to-grain direction

Fig. 3.14 Schematic
diagram of elasto-plastic
material behaviour

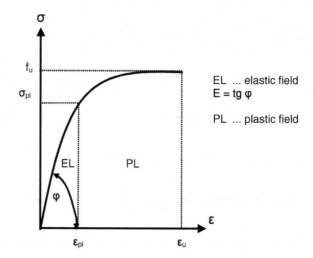

- Timber species
- Irregularities in timber
- The humidity of timber
- The duration of the load actions (rheological phenomenon)
- The temperature.

3.1.6.1 Type-of-Stress-Related Influence

With respect to elasticity, we generally distinguish between the bending modulus
(E_m), compression parallel to the grain modulus ($E_{c,0}$) and tension parallel to the
grain modulus ($E_{t,0}$) in timber elements. The values of $E_{c,0}$ in $E_{t,0}$ are approximately the same only at low specific deformation while they display a progressive
difference after the borderline of proportionality in the compressive stress has been
crossed. Since timber's behaviour in tension as compared to that in compression
displays a substantially higher degree of non-ductility, the modulus of elasticity in
compression slightly surpasses that in tension. In certain cases, their values could
be considered as identical; nevertheless, it is better to use an average effective
value which is defined as follows.

Approximate mathematical note of the relation between the specific deformation (ε) and the stress (σ) as seen below can be applied to most materials:

$$\varepsilon = \alpha \cdot \sigma^n \tag{3.2}$$

In Eq. (3.2), n stands for the material constant having different values
depending on the type of material. Tests on timber proved a linear relation between
both values, which means $n = 1$ and Eq. (3.2) therefore asserts that:

$$\varepsilon = \alpha \cdot \sigma \tag{3.3}$$

If the coefficient α is written separately for the behaviour of timber in compression (α_c) and its behaviour in tension (α_t), the average (arithmetical) value (α_{av}) is:

$$\alpha_{av} = \frac{\alpha_c + \alpha_t}{2} \tag{3.4}$$

If the modulus of elasticity is introduced as $E = 1/\alpha$, its average value is

$$E_{av} = \frac{1}{\alpha_{av}} = \frac{2}{\alpha_c + \alpha_t} = \frac{2 \cdot E_c \cdot E_t}{E_c + E_t} \tag{3.5}$$

As already mentioned in the part discussing the bending strength of timber, the latter is in fact defined as a combination of timber's behaviour in compressive and tensile zones. The average modulus of elasticity can thus be defined as the bending modulus ($E_{av} = E_m$). The following ratio of timber's moduli of elasticity parallel to the grain can be inferred from the above: $E_{c,0} > E_m > E_{t,0}$. Since the values indicated in the area of elasticity up to the point of proportion show little difference with the majority of timber species, we usually assume for the purpose of computation that they are identical.

3.1.6.2 The Influence of the Stress-to-Grain Direction

The behaviour of timber displays, as said beforehand, certain differences based on different directions. According to a fully anisotropic study of timber, its modulus of elasticity parallel to the grain is the highest in the longitudinal direction ($E_{0,L}$), varies from 1/6 to 1/23 $E_{0,L}$ in the radial direction and is the lowest, reaching only 1/11–1/40 of the $E_{0,L}$ value, in the tangent direction. With a simplified orthotropic study, the distinction is to be made between the modulus of elasticity parallel to the grain (E_0) and the substantially lower modulus of elasticity perpendicular to the grain (E_{90}).

3.1.6.3 The Influence of Timber Species

Like timber strength, the modulus of elasticity depends on "the state of compression" of timber, i.e., on the annual ring widths. The narrower these are, the higher are the strength of timber and its modulus of elasticity. As deciduous trees generally have narrower annual ring widths than conifers, the modulus of elasticity at an identical percentage of timber irregularities is estimated to be higher in deciduous trees than in conifers. Growth imperfections can also significantly reduce the value of timber. With a view to a better illustration of the influence exerted by the grain direction and timber species, Table 3.4 lists the modulus of elasticity values of certain typical timber species.

Table 3.4 The modulus of elasticity in different directions for different timber species (by Campredon) (in N/mm^2)

	$E_{0,L}$	$E_{0,R}$	$E_{0,T}$	$E_{90,L}$	$E_{90,R}$	$E_{90,T}$
Softwood conifers (fir, spruce tree)	10,000–14,000	750–1,000	400–500	100	700	700
Semi-hardwood conifers (pine)	12,000–16,000	1,100–1,300	600–800	100	800	1,000–1,200
Softwood deciduous trees (poplar, willow, linden)	9,000–12,000	1,100–1,200	600–800	100	800	1,100–1,200
Semi-hardwood and hardwood deciduous trees (oak, beech, ash)	1,300–1,800	1,500–2,000	800–1,200	400–500	800–1,000	1,200–1,500

The table clearly shows that the modulus of elasticity parallel to the grain is considerably higher than its perpendicular-to-the-grain counterpart. An obvious influencing factor is also the state of timber compression; deviation between softwood conifers and hardwood deciduous trees can reach even 80 %.

A comparison of the timber's modulus of elasticity parallel to the grain with the elasticity modulus of concrete, Table 3.3, exposes its 3 times lower value. Yet, with timber having roughly 5 times lower density than concrete, its effectiveness with respect to the weight of the material used tends to be 2 times lower than that of concrete, which undoubtedly classifies timber structures as lightweight structures.

The modulus of elasticity of timber has a relatively low value in comparison with other construction materials, e.g., 20 times lower than that of steel. Planning and designing timber structures hence call for careful consideration of the potential difficulties in satisfying the requirements of the serviceability limit state and, to a lesser extent, those of the ultimate limit state. Low values of the timber's modulus of elasticity are thus usually a major problem in designing multi-storey timber structures which will be discussed in Sect. 3.4.

3.2 Basic Structural Systems of Timber Construction

As the most easily accessible building material, timber has been offering shelter and dwelling to a man since ancient times. Timber is an organic material whose cycle of use is fully rounded up, from growing in the woods, where it serves as food and shelter to animals, through the phase of being processed into a raw material, semi-manufactures and finished goods to the stage of producing timber biomass from timber waste. Timber buildings have therefore always been an important part of the infrastructure in a number of areas around the world.

At the present times, characterized by specific circumstances in the sphere of climate change, additionally witness an intensive focus of the sciences of civil engineering and architecture on searching for ecological solutions and construction methods that would allow for higher energy efficiency and, consequently, for reduced environmental burdening. Due to the fact that buildings represent one of the largest energy consumers and greenhouse gas emitters (with their share being approximately 40 %) energy-saving strategies related to buildings, such as the use of eco-friendly building materials, reduction in energy demands for heating, cooling and lighting, are strongly recommended. Timber structures thus undoubtedly deserve to be considered as having an important advantage over those made from other construction materials.

Besides the listed ecological benefits, there are currently additional strong arguments in favour of building timber structures, as seen from the structural and technological points of view. Brand new and improved features, being introduced in the early 1980s of the last century, brought about a strong expansion of timber buildings all over the world. The most important changes introduced in timber construction are listed below:

- Transition from on-site construction to prefabrication in factory
- Transition from elementary measures to modular building
- Higher input of glued-laminated timber in construction
- Development from single-panel wall system to macro-panel wall prefabricated system.

A more detailed presentation of the above arguments is to be found further in this section.

Competitive construction fields are aware of the significance attributed to timber-frame structures capable of fulfilling most of the demands of the society and the environment we live in. A list of the foremost arguments in favour of timber-frame homes comprises

- Highly favourable physical properties of the buildings
- Environmental excellence of the built-in materials
- Lower energy consumption in the process of manufacturing built-in materials
- Faster construction
- A more efficient use of indoor space
- Good seismic safety.

Physical properties of the buildings are of utmost importance as good insulation saves the energy needed for heating.

Timber and gypsum as predominant materials use *less energy in the manufacturing process* than elements made of brick, concrete or other prefabricated products. In comparison with other types of buildings, the energy-efficient properties of prefabricated timber buildings are excellent not only because well insulated buildings use less energy for heating, which is environment-friendly, but also due to a comfortable indoor climate of timber-frame houses, Gold and Rubik [1]. Considering the growing importance of energy-efficient building methods, timber construction will play an increasingly important role in the future. The use of timber in construction is gaining ever more support, especially in regions with vast forest resources since it reduces the energy demand for transport if the building material is available from the local area. With respect to all the given facts, timber as a material for load-bearing construction represents a future challenge for residential and public buildings.

Another aspect in favour of timber-frame houses is *faster construction* due to a high degree of elements' prefabrication. Consequently, only on-site construction stage is exposed to weather conditions, and the probability of later claims is lowered.

Furthermore, at *identical outer dimensions,* timber houses *cover up to a* 10 % *larger residential area* than a concrete or masonry wall system. The reason lies in the fact that in spite of a smaller wall thickness, timber houses have better thermal properties than those built by using conventional brick or concrete construction systems. All in all, lower maintenance costs, high thermal efficiency and lower probability of constructional failure make it easier for investors to choose this option.

Timber structure is commonly associated with lightweight construction and has relatively high ductility whose degree is additionally raised through the flexibility of mechanical fasteners in the connections between timber elements, all of which results in timber structures maintaining a good performance, particularly when exposed to *wind* or *earthquake forces*.

To sum up, numerous advantages have contributed to an ever larger proportion of prefabricated timber construction worldwide. The following comparative

Fig. 3.15 Overview of the basic structural systems in timber construction; (**a**) log construction, (**b**) solid timber construction, (**c**) timber-frame construction, (**d**) frame construction, (**e**) platform-frame construction, (**f**) panel construction; photograph by F. Kager

numbers show the percentage of newly erected prefabricated timber residential buildings in different parts of the world: Canada 95 %, USA 65 %, Japan 50 %, Scandinavia 70 %, Great Britain 10 % (Scotland 50 %), Germany 7 % (Bavaria 30 %), Austria 8 %, Czech Republic 2 % and South Europe up to 3 %, Lokaj [2]. Evidently, there are currently substantial differences in the global expansion of prefabricated timber structures.

3.2.1 Short Overview of Basic Structural Systems

Selecting a timber construction system depends primarily on architectural demands, with the orientation, location and the purpose of a building being of no lesser importance. Prefabricated timber construction systems differ from each other in the appearance of the structure and in the approach to planning and designing a particular system. As presented in Kolb [3], timber houses can be classified into six major structural systems:

- Log construction (Fig. 3.15a)
- Solid timber construction (Fig. 3.15b)
- Timber-frame construction (Fig. 3.15c)

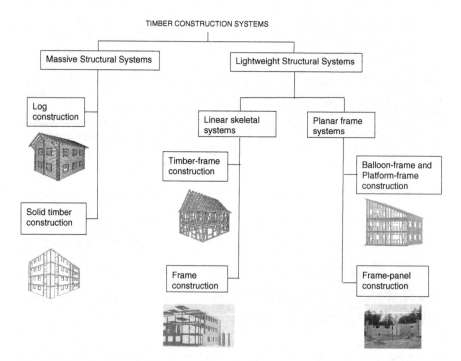

Fig. 3.16 Classification of timber structural systems according to their load-bearing function

- Frame construction (Fig. 3.15d)
- Balloon- and platform-frame construction (Fig. 3.15e)
- Panel construction (Fig. 3.15f).

Log construction and solid timber construction can also be classified as *massive structural systems* since all load-bearing elements consist of solid elements. Other construction systems shown in pictures (c–f) consist of timber-frame-bearing elements and are therefore classified as *lightweight structural systems*. According to the load-bearing function, they can be subdivided into classical *linear skeletal systems* where all the loads are transmitted via linear bearing elements and *planar frame systems* where sheathing boards take over the horizontal loads. The classification of structural systems based on their load-bearing function is schematically presented in Fig. 3.16.

The following subsections offer a brief description of the above structural systems based on the main characteristics of the load-bearing behaviour of structural elements. Log and timber-frame construction systems are typical of traditional types of timber houses whose predominance in the past, especially in the countries with huge wooded areas and a strong timber tradition and industry, is now giving way to construction systems currently dominating the market, i.e., mostly to panel construction, solid timber construction and frame construction.

However, the main focus will be laid on the panel construction system whose detailed analysis in combination with the glazing is the subject matter of Chap. 4. As the panel construction system consists of timber-frame elements (studs and girders) and sheathing boards (panels), the term *timber-frame-panel construction* will be used a substitute in further parts of the book.

3.2.2 Massive Timber Structural Systems

3.2.2.1 Log Construction

Log construction is the most traditional type of timber construction, used in many countries in the world, especially in areas with cold climate conditions. Scandinavia, for example, is home not only to old residential buildings where log construction system was applied (see Fig. 3.17), but also to other structures, such as churches, towers. Furthermore, in the Alps and the mountainous regions of Central Europe, log construction usually plays an important role, especially for local inhabitants' houses.

Log construction is classified as a massive timber structural system since its load-bearing elements consist of solid elements. Log construction is the most massive and usually the most expensive type of timber construction. The building's envelope consists of a single leaf of horizontally stacked timber members (Fig. 3.18a) which need to perform a triple function, that of cladding, space enclosing and load bearing. Stability is achieved through the friction resistance in

Fig. 3.17 Traditional type of
log construction

the bed joints, which makes the solid timber wall act as a plate, and through the cogged joints between the timber members at the corners, Deplazes [4]. The old-type system (Fig. 3.18a) usually requires no mechanical fasteners. A new type of connection, also called the "dovetail" joint, is mostly used in modern log buildings (Fig. 3.18b).

The traditional old types of log structures can seldom satisfy modern standards of the comfort of living or those of the energy efficiency. For example, the thermal transmittance of a 30-cm-thick wall made of solid conifer timber with no insulation amounts to $U_{wall} = 0.40\,\text{W}/\text{m}^2\text{K}$, which is a higher value than required by modern standards. Such construction form is thus no longer as widespread as in it used to be in the past.

(a)   **(b)**

Fig. 3.18 Types of connections in log construction; (**a**) the old type, (**b**) the new type

3.2.2.2 Solid Timber Construction

Solid timber construction (Fig. 3.15b), as one of the most widely used types of timber construction today, is a strong competitor to the frame-panel structural system, described in Sect. 3.2.3.4. Both types are prefabricated systems (Fig. 3.19) characterized by a very short platform erection time, which justifies the slogan "only two men to build a house".

Nevertheless, from a structural point of view, there is an important difference between the above-mentioned systems. Due to its timber-frame bearing construction, the frame-panel system belongs to "Lightweight Timber Construction" (LTC), while the solid timber system with its solid panel composition falls into the category of "Massive timber construction" (MTC). Massive timber construction is more expensive, but it generally has higher horizontal stiffness and load-bearing capacity than the frame-panel construction and is therefore more appropriate for multi-storey timber buildings. In addition, with the MTC system, no vapour barrier is generally needed, which accounts for a higher heat storage capacity in comparison with LWC systems. The above facts could also be important in using timber and glass together with a view to achieving higher solar gains by inserting an optimal combination size of timber and glazing in the structure, cf. Sect. 4.2. All the listed facts have to be considered when choosing the most appropriate type of timber building.

There are many different types of contemporary massive solid timber constructions found all over the world. They can be subdivided into *massive panel structural systems* where glued panel floor and wall elements resist all the loads in the structure, and *modular building block systems* consisting of hand-size modular blocks.

Fig. 3.19 Platform-type erection in the massive panel timber construction, photograph by F. Kager

Massive Panel System

There are many different types of massive panel structural floor and wall elements used as modular elements, usually built in standard dimensions, shown in Table 3.4.

An important and most frequently occurring massive construction type today is the massive panel system using wall and floor elements made from *Cross-Laminated Timber* (*CLT*) *panels*. In Switzerland, for example, the popularity goes to Leno solid panels. These are solid cross-laminated timber panels made from three to eleven fir plies glued together crosswise, Fig. 3.20. The resulting homogeneous, dimensionally stable and rigid component can be produced in sizes up to 4.8 × 20 m. Available thicknesses depending on the number of plies range from 50 to 300 mm, Deplazes [4].

Wall elements in massive panel timber buildings consist of cross-laminated timber boards, known as "cross-laminated timber—CLT" or "Kreuzlagenholz— KLH" or "X-lam" (Fig. 3.21). Basic material for the production of CLT elements is sawmill-boards, whose quality is best if they are cut from the outer zones of the log. Such sideboards, which are not considered as particularly profit-making items by the millers, generally have excellent mechanical properties relating to stiffness and strength. The width of the boards needed for the production of CLT elements is usually 80–240 mm, with their thickness ranging from 10 to 45 and up to 100 mm (depending on the producer). The width to thickness ratio should be defined as $b:d = 4:1$. Timber species currently processed belong to conifers (e.g. fir, spruce fir, pine), which does not exclude deciduous trees (e.g. ash, beech) from being used in the future, Augustin [5].

The crossing of board layers results in good load distribution properties in both directions. The width of a wall element depends on the number of layers: 3, 5 or even 7. A typical three-layered wall element is 90–94 mm wide. Owing to the gluing of longitudinal and transverse layers, the "activity" of timber is reduced to a negligible degree.

Cross-laminated timber is a contemporary building material having more uniform and better mechanical properties than solid timber. It therefore represents not only an architectural challenge but also an important trend in the construction of

Fig. 3.20 LenoTec product with 5 spruce fir plies glued together crosswise, t = 135 mm, 5 plies

Fig. 3.21 Composition of a
bearing timber construction
in cross-laminated panels

modern, energy-efficient and seismic resistant single-family prefabricated houses
and multi-storey prefabricated residential timber buildings. CLT is also suitable
for offices, industrial and commercial buildings. On-site assembly methods of fully
prefabricated wall elements are similar to those applied with timber-frame
buildings and take an almost equal amount of time.

The size and form of CLT elements are defined by restrictions concerning
production, transport and assembly. The existing standard dimensions of planar
and single-curved elements are set at a length, width and thickness of 16.5, 3.0 m
and up to 0.5 m, respectively. Longer elements (of up to 30 m) can be assembled
by means of general finger joints. The thickness of lamellas in curved CLT elements has to be adjusted to the curvature, Augustin [5]. In general, the use of CLT
elements is restricted to service classes 1 and 2 according to Eurocode 5 [6].

Insulation in cross-laminated timber wall elements is placed on the external
sides of the bearing timber construction, as shown in Fig. 3.22. This is one of the

Fig. 3.22   CLT external wall
elements; (**a**) normal type,
(**b**) thermally improved
element with additional
insulation placed in the
timber substructure

(**a**)

(**b**)

most significant disadvantages of CLT wall elements in comparison with frame-panel wall elements. Producers usually offer a variety of massive panel elements of different thicknesses (see Table 3.5) which can be combined with different thicknesses of insulating materials to form wall elements having different thermal properties, according to the chosen type of the external wall element—that of low energy or passive standard. According to the data presented in Table 3.5, the composition of wall elements differs in the construction thickness of the load-bearing CLT structure as well as in that of insulation placed in the additional timber substructure whose usual thickness of 60 mm is by no means not the only dimension used in production.

There are also some other types of solid panels wall and floor elements on the market which can be treated as box elements. For example, Lignotrend consists of three to seven cross-banded softwood plies with an average total thickness of 125 mm, where gaps of several centimetres separate individual pieces of the inner plies. The raw material is solely side boards or softwood. We can distinguish between elements opened on both sides (Fig. 3.23a), those closed on both sides (Fig. 3.23b) or just on one side (Fig. 3.23c). Wall elements are supplied in widths up to 625 mm, Deplazes [4] and Deplazes et al. [7].

Typical characteristics of selected prefabricated solid timber panels made of CLT and box elements are listed in Table 3.5.

Modular Building Block System

One of the newest types of massive timber structural system is the modular building block system. Base load-bearing elements are small-format, factory-made

Table 3.5 Typical dimensions of some solid timber panels most commonly used in the production

Producer	Number of plies	Thickness (mm)	Planar max. dimensions (mm × mm)	Type
Binderholz	3	78	3,000 × 16,500	CLT
Binderholz	3	90	3,000 × 16,500	CLT
Binderholz	5	100	3,000 × 16,500	CLT
KLH	3	57	3,000 × 16,500	CLT
KLH	3	72	3,000 × 16,500	CLT
KLH	3	94	3,000 × 16,500	CLT
KLH	5	95	3,000 × 16,500	CLT
KLH	5	128	3,000 × 16,500	CLT
Leno	3	81	4,800 × 20,000	CLT
Leno	5	135	4,800 × 20,000	CLT
Leno	7	216	4,800 × 20,000	CLT
Bresta		80–260	2,800 × 12,000	Edge-fixed elements
Ligno-Swiss	/	100–360	3,100 × 15,000	Box element
Schuler-Blockholz	/	18–500	3,000 × 9,000	Ribbed element
Ruwa Holzbau	/	50–200	200 × 20,000	Block system

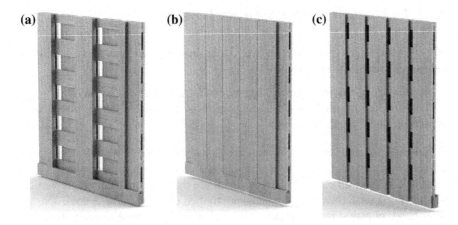

Fig. 3.23 Three different types of Lignotrend panel wall production; open on both sides (**a**), closed on both sides (**b**), closed on one side (**c**)

modules of solid timber, usually in the form of building blocks. Modules are built up in a type of masonry bond following modular dimensions to form load-bearing internal and external walls (Fig. 3.24). Standardized bottom plates, modules, top plates and opening trimmers for standard doors and windows result in a coordinated building system, Kolb [3]. Owing to the modular form, this construction system with hand-size block technology has undergone considerable development and has already become quite widespread.

Fig. 3.24 Modular building block system

3.2.3 Lightweight Timber Structural Systems

We can distinguish between two main load-bearing structural systems within lightweight structural systems:

- Linear skeletal systems (timber-frame construction, frame construction)
- Planar frame systems (balloon-frame construction, platform-frame construction, frame-panel construction).

The essential difference between the two systems is seen in the following facts: the sheathing boards or infill material in linear systems do not contribute to the resistance of the wall elements. All the loads are therefore transmitted via vertical (studs), horizontal (beams) and diagonal timber elements (Fig. 3.25a). On the other hand, in planar frame systems, the studs are placed close to each other thus allowing the boards to transmit the horizontal loads and be consequently treated as resisting elements (Fig. 3.25b).

3.2.3.1 Timber-Frame Construction

Timber-frame construction, seldom used today, is one of the traditional forms of lightweight timber structures. Timber-frame buildings found in many countries of Northern, Central and Eastern Europe exhibit numerous regional characteristics. This type of structural system commonly developed in regions where timber was not available in sufficient quantities required for log construction. Until the middle of the nineteenth century, most timber-frame structures had a visible load-bearing framework (and the infill panels). Bricks or clay bricks were usually installed to fill the spaces between timber elements forming the frame, where timber was based on a relatively small module with diagonal braces in the wall plane (Fig. 3.26). »Builders believed that this gave their buildings an improved fire resistance, but

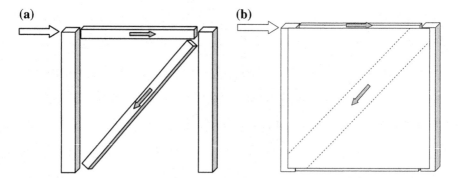

Fig. 3.25 Horizontal force distribution in (**a**) linear skeletal system, (**b**) planar frame system

Fig. 3.26 Rural types of old timber-frame construction with a visible load-bearing frame

there was also a desire to give an "urban" look to timber construction, which was regarded as a "rural" form of construction«, Kolb [3].

Timber-frame construction shows first signs of prefabrication (Fig. 3.27). Assembly of individual elements takes place on-site, storey by storey, with the horizontal loads transmitted via beams and diagonals directly to the base. The spacing between individual timber studs depends on the load-bearing capacity of timber sections which (in times before the industrialization) used to be cut to the required size by using simple tools. The vertical loads are transferred directly via contact faces between various timber members, Deplazes [4].

3.2.3.2 Frame Construction

From a structural point of view, frame construction could be regarded as the second type of linear skeletal systems where all loads are transmitted via vertical (studs), horizontal (beams) and diagonal elements (Fig. 3.25a). Stability is achieved by the inclusion of diagonal elements which usually extend through all the storeys. The load-bearing structure thus functions independently of the

Fig. 3.27 First signs of prefabrication in timber-frame construction

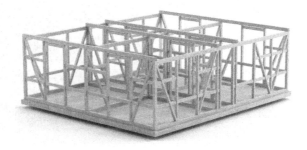

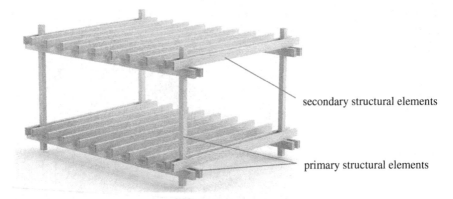

secondary structural elements

primary structural elements

Fig. 3.28 Primary and secondary structural elements in frame construction

enclosing elements, such as façades or sheathing boards. The structure is therefore divided into two load-bearing substructures (Fig. 3.28):

- The primary structure
- The secondary structure.

The *secondary structure*, consisting of timber floor or roof elements placed within small distances from each other, transfers the loads from the roof, the suspended floors and the walls to the main beams, which calls for relatively high floor elements.

The *primary structure* consists of the load-bearing columns and main beams, usually of rectangular cross sections. It carries the loads from the secondary structure and transfers these as concentrated loads to the foundations.

An old type of frame structure is the so-called post and beam construction (Fig. 3.29) where timber columns are placed within relatively large distances. Continuous beams of considerably large spans and cross-sectional dimensions are

Fig. 3.29 Post and beam structural system

supported by one-level high columns, with additional corner diagonal elements being usually inserted in order to strengthen the flexible connection between both load-bearing elements and to assure the horizontal stability of the building. This type of frame construction is probably one of the oldest structural forms besides the log construction.

The growing importance of timber construction in the fields of multi-storey and large-volume structures has helped modern frame construction, which is probably the most attractive current form of timber construction, to take on a new role in creating one of the most important and attractive types of timber structures of the present times. Although the mentioned classification of the elements to perform the secondary and primary load-bearing functions is not very economic in terms of material consumption, it does have its benefits since it leads to substantial flexibility in terms of the internal layout and design of the façade. Moreover, it opens a possibility of using longer spans with fewer internal columns than in other construction systems, which results in greater freedom of designing the interior layout. In addition to houses and other residential and public buildings, frame construction is suitable for offices, industrial and commercial buildings, Kolb [3].

Since the load-bearing functions in the frame system belong to the timber-frame elements and diagonals and due to longer spans, additional requirements for timber members arise. For instance, there is a preference to use glued-laminated timber for timber frames instead of the classic sawn timber. On the other hand, the sheathings and the timber frame do not have any resisting function which is performed by the diagonals. Consequently, the walls with a higher number and an enlarged size of openings can be used (Fig. 3.30).

Frame structure is probably the most advanced structural type of timber buildings today. Satisfying most of the energy-efficiency demands, whose detailed description and explanation are part of Chap. 2, the frame structure is ideal for combining timber and glass, which can lead to designing highly attractive buildings. The timber frame is a fully resisting part of the building where the two materials are combined, while the glazing has no load-bearing function but assures solar gains and the transparency of the building. In addition to an attractive design, such buildings offer a most comfortable indoor living climate.

Fig. 3.30 Frame structural system—walls with an enlarged size of openings can be used in timber buildings

(a)   **(b)**

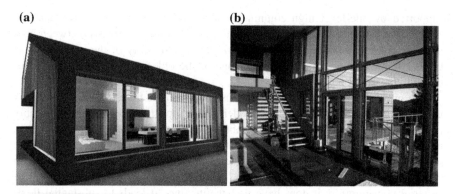

Fig. 3.31 Large glazing surfaces in timber buildings: (**a**) one-storey house designed in a study workshop Timber low energy house and (**b**) single-family house in Hohe Wand, Austria, designed and photograph by Architekturbüro Reinberg ZT GmbH Vienna

Furthermore, two vitally important facts, which will be further discussed in Sects. 3.4 and 4.4 need to be stressed. Firstly, owing to a rather low impact of the horizontal loads in one-storey houses, there is usually no need for additional diagonals to assure the horizontal stability of the structure (Fig. 3.31a). On the other hand, a strong impact of the horizontal loads in multi-storey buildings calls for the insertion of diagonal elements, which remain visible, to assure the horizontal stability of the building (Fig. 3.31b).

3.2.3.3 Balloon-Frame and Platform-Frame Construction

As far as the statics is concerned there is almost no difference between the balloon-frame and the platform-frame constructions. In both structural types, the closely spaced studs with squared sections, usually of standard sizes, are placed within a constant distance range. Connecting the boards to the timber studs takes place on-site after the erection of the timber structures. As far as the force transmission is concerned, both construction types can be treated as planar frame systems since the sheathing boards have an important load-bearing function in transmitting the horizontal load (Fig. 3.25b) and assure the horizontal stability of the structure.

Yet, from the technological viewpoint, there is a significant distinction between the two construction types. In the *balloon-frame construction* (Fig. 3.32a), the function of the main load-bearing part is taken by the frame composed of beams and pillars following a continuous bottom-to-roof pattern. The joists of suspended floors are supported by a horizontal binder inserted into notches cut in the vertical studs. All connections are usually fastened with nails. The maximum height of pillars is limited to approximately 8 or 10 m, which makes the balloon-frame construction suitable solely for one or two-storey buildings. This construction system is particularly well known in North America. In Europe, the *timber stud*

(a)   **(b)**

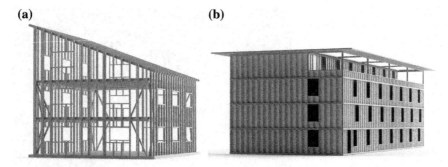

Fig. 3.32 Balloon-frame construction with continuous vertical studs (**a**) and platform-frame construction with storey-by-storey assembly (**b**)

construction with a slightly lesser degree of prefabrication is the equivalent of the American balloon frame.

In the *platform-frame construction* (Fig. 3.32b), the load-bearing timber frame consists of storey-high preassembled studs with square sections. Individual load-bearing frame elements are assembled at the prefabrication stage and transported to the construction site as self-contained elements. In the following stage, the boards are connected to the timber frame to assure the horizontal stability of the building. The construction of platforms is a step-by-step procedure.

The main advantage of both structural types is a considerably low weight of the building. However, Europe has witnessed a major replacement of the two construction types with the frame-panel construction where prefabrication encompasses a more significant part of the construction process.

3.2.3.4 Frame-Panel Construction

The frame-panel system originates from the Scandinavian-American construction methods, i.e., balloon-frame and platform-frame construction types whose assembly works take place on-site. Advantages of the frame-panel construction system over the above-mentioned traditional timber-frame construction systems were first noticed at the beginning of the 1980s of the previous century and made a significant contribution to the development of such timber construction. The benefits to be pointed out lie in factory prefabrication (Fig. 3.33) assuring the so-called ideal weather conditions in addition to constant supervision over construction works and the built-in materials. Another asset is the subsequently faster assembly process as the ready-made elements are crane-lifted to be erected onto the foundation platform, adjusted and screw-fastened.

Besides having good constructional characteristics, the panel construction system offers a series of additional advantages among which indoor climate of timber buildings is of major importance since the built-in materials are natural, people-friendly and demand very little energy for their provision. The panel

Fig. 3.33 Production of
load-bearing wall elements in
a factory

construction system's priority over massive masonry construction is seen in thinner wall elements, which can result into an increase in the usable floor area by 10 %, with the total surface area of a building being the same.

Furthermore, the transition from the single-panel construction system (Fig. 3.34a) to the macro-panel construction system (Fig. 3.34b) means an even higher assembly time reduction and higher stiffness of the entire structure due to a lesser number of joints. The wall elements with a total length of up to 12.5 m are now entirely produced in a factory (Fig. 3.33).

The basic vertical load-bearing elements in the frame-panel construction, which is nowadays typical of Central Europe, are the panel walls consisting of the load-bearing timber frame and the sheathing boards, while the horizontal floor load-bearing function goes to slabs made of the floor beams and the load-bearing wood-based sheathing boards connected to the upper side of the floor beams. The construction is a systematic storey-by-storey creation of a building; after the walls have been erected, the floor platform for the next level is built. The usefulness along with the popularity of this system in the case of multi-storey buildings has generated a growing interest in its application around the world.

Prefabricated timber-frame walls functioning as the main vertical bearing capacity elements, whose single panel typical dimensions have a width of $b = 1,250$ mm and a height of $h = 2,500–3,100$ mm, are composed of a timber frame and sheets of board-material fixed by mechanical fasteners to both sides of the timber frame (Fig. 3.34). There are many types of panel sheet products available which may have a certain level of structural capacity such as wood-based materials (plywood, oriented strand board, hardboard, particleboard, etc., or fibre-plaster boards). Their use originally started in Germany and these products have recently become the most frequently used types of board-materials in Central Europe. A thermo-insulation material, whose thickness depends on the type of external wall, is inserted between the timber studs and girders. The sheathing boards on both sides of the wall can be covered with a 12.5-mm gypsum–cardboard.

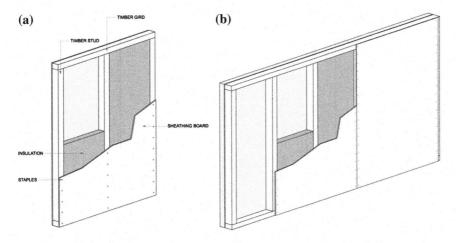

Fig. 3.34 Single-panel construction system (**a**) and macro-panel construction system (**b**)

In the frame-panel construction, the entire wall assemblies, including windows, doors and installations, are fully constructed in a horizontal plane (Fig. 3.35a) in the factory and then put in vertical position (Fig. 3.35b).

Producers usually offer a variety of composite construction wall elements of different types related to energy efficiency. In Central Europe, the most common type of production of the external walls of low energy-efficiency standard is the timber frame with a thickness of 160 mm. The whole composition, labelled as TF-1, is presented in Table 3.6. The improved type with an additional timber substructure and insulation of 60 mm is labelled as TF-2. The passive type of the external wall is labelled as TF-3. The cross sections are graphically presented in Fig. 3.36a, b and c.

A difference in the insulation properties of the wall elements, such as the insulation thickness, the number of insulating layers or the insulation type can be

Fig. 3.35 The walls with openings are constructed in horizontal position (**a**) and then put in vertical position (**b**)

Table 3.6 Composition of three typical external wall elements

TF-1		TF-2		TF-3	
Material	Thickness (mm)	Material	Thickness (mm)	Material	Thickness (mm)
Rough coating	6	Rough coating	6	Rough coating	9
Thermal insulation: EPS[b]	100	Thermal insulation: MW[a]	100	Thermal insulation: wood fibreboard	60
Gypsum fibreboard	15	Gypsum fibreboard	15	/	/
Thermal insulation: MW[a] between timber frame	160	Thermal insulation: MW[a] between timber frame	160	Thermal insulation: CF[c] between timber frame	360
Vapour barrier	0.2	Vapour retarder	0.2	OSB	15
/	/	Thermal insulation: MW[a] between timber substructure	60	/	/
Gypsum fibreboard	15	Gypsum fibreboard	15	/	/
Gypsum fibreboard	10	Gypsum fibreboard	10	Gypsum plasterboard	12.5
Total thickness [mm]	306.2	Total thickness [mm]	366.2	Total thickness [mm]	456.5
U_{wall}–value (W/m²K)	0.164	U_{wall}–value (W/m²K)	0.137	U_{wall}–value (W/m²K)	0.102

[a] Mineral wool, [b] expanded polystyrene foam, [c] cellulose fibre

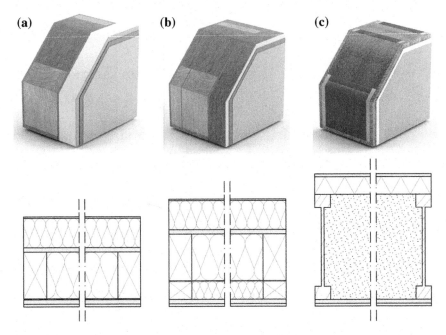

Fig. 3.36 Composition of three typical timber-frame external macro-panel wall elements; low-energy system (**a**) improved low-energy system with installation layer (**b**) passive system (**c**)

observed in the above table. Consequently, the thermal transmittance of the three wall elements (U_{wall}–value) differs as well.

As the erection of buildings constructed in the frame-panel structural system follows the platform technology, the height of the wall elements is equal to that of the storey. In residential buildings, the height usually ranges from 2,500 to 2,700 mm. On the other hand, the relevant height in public and commercial buildings may exceed the previously mentioned numbers. The maximal height of the prefabricated wall assembly constructed in the macro-panel system is limited—it depends on transportation requirements and consequently on the total length of the wall assembly. Maximal dimensions given in Table 3.7 and are graphically presented in Fig. 3.37.

The wall elements are usually prefabricated in standard modular dimensions, with commonly used 625 mm spans between the timber studs (Fig. 3.38). However, to satisfy architectural design requirements for special forms of buildings,

Table 3.7 Maximal dimensions of prefabricated wall assemblies

Maximal height (mm)	Maximal length (mm)
2,850	12,000–12,500
3,050	9,000

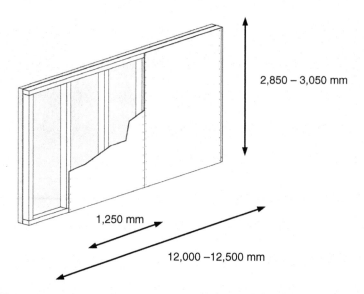

Fig. 3.37 Maximal dimensions of prefabricated wall assemblies

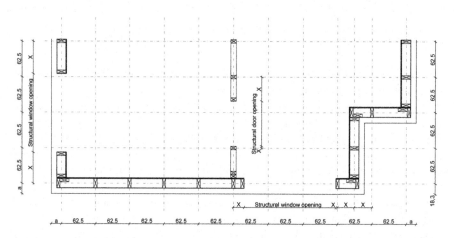

Fig. 3.38 Modular dimensions in prefabricated frame-panel wall elements, dimensions in cm

exceptions can be adopted in, e.g., window and door openings, fixed glazing areas. When such exceptions take place, the wall elements next to the opening or those at the end of the whole assembly are usually cut to get the required dimension.

3.3 Design Computational Models

The main elements transferring forces to the foundation in prefabricated timber-frame-panel structures are timber-frame-panel walls which belong to the single-panel system (Fig. 3.34a) or to the macro-panel system (Fig. 3.34b).

The vertical load (dead load, live load, snow, etc.) impact is calculated as axial compression parallel to the grain in the timber-frame studs. The above impact, however, is not the subject of the present publication.

The focus of this section is on the *horizontal load* (wind, earthquake) distribution to the load-bearing elements of the frame-panel construction. In a structural system consisting of vertical prefabricated walls and prefabricated slabs acting as rigid diaphragms, horizontal displacements at the top of a storey due to horizontal loads are constant along a wall assembly. The horizontal load is thus distributed to wall assemblies and further to individual wall elements according to their stiffness ratios (Fig. 3.39).

The platform-frame and balloon-frame constructions have a similar load distribution. A sole technological distinction is that of the elements not being prefabricated but usually erected on-site.

In the past, many computational models were suggested to calculate the horizontal load distribution in timber-frame-panel wall assemblies. As described in Breyer et al. [8], Faherty and Williamson [9], Thelanderson and Larsen [10], Schulze [11] and Eurocode 5 [6], the most common way to calculate structural behaviour in the timber-frame-panel wall assembly under the horizontal load is to assume that each floor platform is rigid, with each timber-frame wall being a vertical cantilever beam fixed at the bottom and free to deflect at the top (Fig. 3.40). Both supports approximate the influence of the neighbouring panel walls and assure a boundary condition for the wall in question. The supports can be considered rigid (Fig. 3.40), as seen in Faherty and Williamson [9] and Eurocode 5 [6], or flexible, which is more realistic in addition to being more accurate and will be further presented and discussed in Sect. 3.3.1.

Each wall assembly at individual levels consists of separate wall segments acting as individual cantilevers, where every segment is determined with the width b of the sheathing board (usually 1,250 mm). As mentioned beforehand, if the lateral forces acting at the top of the element are considered to be uniformly distributed to each segment, the horizontal force acting on a single wall element can be calculated as $F_H = F_{H,\text{tot}}/n$. Since only the segments with the full wall height having no window and door openings are usually taken into account for the calculation, the value n represents the number of single-panel wall elements without any openings (Fig. 3.40).

Many design computational models have been proposed in order to analyse and predict the racking resistance of the timber-frame wall subjected to the lateral loads. According to the accuracy of approach, they can be subdivided into the following groups:

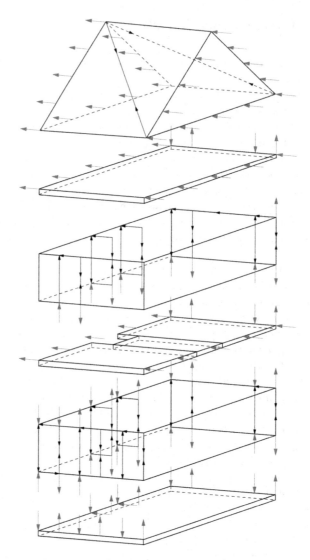

Fig. 3.39 Force distribution under the horizontal load in the timber-frame-panel building, adopted from Kolb [3]

- FEM models
- 2D-braced frame models with fictive diagonals
- Semi-analytical simplified shear models
- Semi-analytical simplified composite beam models.

A good mathematical model is crucial for quick and accurate calculations in the case of architecturally more complex and irregularly shaped buildings with a

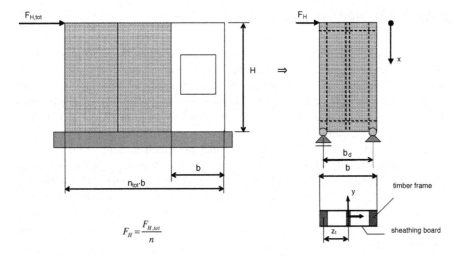

Fig. 3.40 Static design for a wall assembly on individual storey

substantial number of wall elements. Mathematical models differ in their accuracy and simplicity and consequently also in the level of suitability for application in practice. The major problem of the timber-frame wall elements is mechanical fasteners which are unable to provide a fully rigid connection. The flexibility of fasteners in the connecting plane between the timber frame and the boards should therefore be taken into account in modelling such wall elements.

The most accurate way of modelling timber-frame walls is the *Finite Element Method* (*FEM*) where the simulation process replaces the timber frame with line elements, the sheathing board with shell elements and the staples with spring elements. Drawbacks of the above approach are its time-consuming nature and unprofitability in addition to its requirement for using specific and more expensive programs. FEM is more suitable for scientific purposes, while in practice simpler and cheaper programs with no special modulus for the described simulation are seen as more adequate and should be therefore given preference.

In civil engineering practice, *two-dimensional* (*2D*)-*braced frames* are frequently used. Their advantage over other models is the possibility of using simpler and cheaper software with a relatively fast modelling stage. Moreover, with properly defined dimensions of fictive diagonals and supports, braced frames offer a very good approximation the real situation.

Engineering practice also encompasses more useful and *simplified methods using a semi-analytical approach*. No specific software is needed for the calculation of the results, as hand calculation suffices but they require some important and strong simplifications which call for careful consideration. A thorough presentation of the above methods follows in Sects. 3.3.3 and 3.3.4.

3.3.1 FEM Models

The process of modelling timber-frame wall elements under the horizontal load by using the finite element method is the most accurate and the most complex approach. Applying the FEM allows for the influence of openings and fixed glazing on the lateral resistance and stiffness of the wall elements to be taken into consideration, which is vital for our further study presented in Sect. 4.4. Yet, with an opening in the wall element, frame-type instead of beam-type behaviour is expected and the semi-analytical beam model presented in Sect. 3.3.4 is consequently not applicable. A numerical study using the FEM analysis, presented at the end of this subsection, was therefore performed by Kozem Šilih [12]. It encompassed the wall elements along with the influence of any possible window or door openings. The FEM modelling of the timber-frame walls calls for additional consideration of the following items which exert significant influence on the horizontal resistance and stiffness:

- Flexibility of mechanical fasteners connecting the boards and the timber frame
- The appearance of cracks in the tensile area of the fibre-plaster boards—when fibre-plaster sheathing boards are used
- The influence of potential flexibility of the surrounding wall elements
- The influence of openings or that of fixed glazing in the wall element
- The influence of inserted steel or carbon diagonals in the case of strengthening the wall elements.

The wall elements studied were modelled and analysed by using the commercial FEM computer software SAP 2000 (2010), which is more thoroughly presented in Kozem Šilih [12] and Kozem Šilih and M. Premrov [13], while the current subsection focuses only on the most important facts of the modelling.

Behaviour of the analysed wall elements is largely tied to the properties of the sheathing material, i.e., to its possible low tensile strength in the case of fibre-plaster boards, and the consequent *appearance of cracks*. The sheathing material is therefore modelled as linear elastic in compression, while in tension a stress drop occurs when the characteristic tensional resistance (f_{bt}) is reached, corresponding to the model of brittle failure. The sheathing boards are modelled by using the nonlinear shell elements. According to the definition of the applied shell elements, nonlinear behaviour is accounted for by two decoupled 1D models: one for the horizontal direction, and one for the vertical direction.

The *timber-frame material* is considered as an isotropic elastic material (with the modulus of elasticity $E_{0,\text{mean}}$), and the elements of the timber frame are modelled as the simple plane-stress elements. Due to their geometry, the timber members behave predominantly as beam elements with the normal stresses acting parallel to the grain, while the normal stresses in the perpendicular direction are negligible.

Mechanical fasteners connecting the timber frame and the sheathing boards also require an in-depth discussion. The fasteners are modelled using the nonlinear

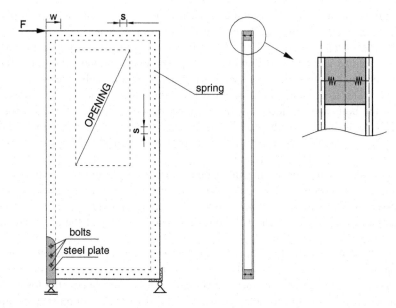

Fig. 3.41 Geometry of the numerical model of the wall element (*left* elevation, *right* cross section with the springs simulating the fasteners)

link elements (springs) with a multi-linear elasto-plastic force–displacement relation (Fig. 3.41). In order to take into account the general in-plane link between the connected elements, two perpendicular uncoupled 1D springs are used for each fastener.

According to Eurocode 5 [6], the shear stiffness (i.e. the slip modulus K) of the fasteners at a specific point depends on the current value of the shear force (V_z). The corresponding three-linear diagram (Fig. 3.42) for the slip modulus K is therefore introduced into the model. The presented numerical model encompasses

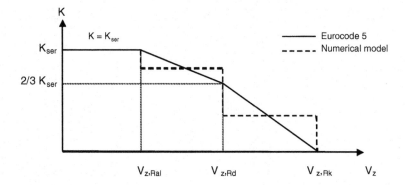

Fig. 3.42 Three-linear diagram for the slip modulus (K) of the fasteners (staples)

a simplified shear force-slip modulus with a constant value of the slip modulus through each phase, indicated by the dashed line in the figure.

The *supporting conditions* (Fig. 3.41) follow the setup for experimental tests and are defined according to Eurocode 5 [6]. The tensile support (bottom left) is arranged using three bolts and two steel plates (one on each side of the wall element). The steel plates are connected to a rigid steel frame. In the numerical model, the bolts are considered as linear elastic spring supports with the stiffness equal to that of the slip modulus (K_{ser}) for bolts. The steel plates are not included in the numerical model. The compressive support (bottom right) is modelled using rigid point supports (Fig. 3.41).

Our further study concerning the influence of fixed glazing on the horizontal resistance and stiffness of the wall follows in Sect. 4.3. Yet, at the present stage, the first step is to focus on the *influence of window or door openings*, with no sheathing material in the area of the opening. The geometry of the analysed wall elements is shown in Fig. 3.43.

The elements consist of a timber frame and are coated on both sides with single fibre-plaster boards having a thickness of $t = 1.5$ cm. The cross-sectional dimensions of the elements of the timber frame amount to 8/9 cm (the lower and the upper beam), 9/9 cm (the side vertical studs) and 4.4/9 cm (all internal beams). The sheathing boards are connected to the timber frame with steel staples. An average intermediate distance between the staples amounts to 7.5 cm, with the exception of the intermediate stud in element G2 and the intermediate studs bellow the openings in elements O1 and O2 where the distance between the staples is 15 cm. Details concerning timber in fibre-plaster materials used are given in Table 3.8.

In view of a better understanding, the numerical results obtained for the wall elements with openings (O1, O2, OM and OV) are compared to those without any

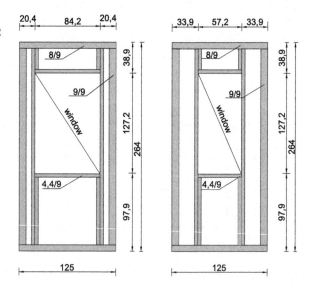

Fig. 3.43 The geometry of the elements O1 (*left*) and O2 (*right*) (the unit of measurement = cm)

Table 3.8 Material properties of the used elements (in N/mm^2)

Timber (C22)				Fibre-plaster boards			
$E_{0,mean}$	$f_{t,0,k}$	$f_{c,0,k}$	$f_{m,k}$	E_b	G_b	f_{bt}	f_{bc}
10,000	13	20	22	3,000	1,200	2.5	20

Table 3.9 Comparison of the numerical results

Type of the wall element	Horizontal resistance ratio		Horizontal stiffness ratio
	F_{cr}	F_u	K_i
G2	1.00	1.00	1.00
OM	0.88	0.89	0.62
O2	0.52	0.59	0.29
O1	0.28	0.39	0.15
OV	0.19	0.27	0.04

openings (G2) without any openings (Table 3.9). In addition to the values of the force at which the first crack is observed in the fibre-plaster board (F_{cr}) and the ultimate capacity—total failure of the test sample (F_u), two additional quantities are dealt with:

- The initial stiffness K_i, defined as the ratio between F_{cr} and the corresponding displacement w_{cr} at the formation of the first crack; K_i thus corresponds to the average stiffness in the range of elastic behaviour (i.e. the region of reversible deformations).
- The value of the force (F_{wt}) at which the defined target displacement $w_t = H/500 = 13.20$ mm is reached; the target displacements are defined with respect to the regulations on the limitations of inter-storey drifts set by the European code for the design of earthquake-resistant structures, Eurocode 8 [14].

The results for the failure force (F_{wt}) at the ultimate displacement (w_t) according to Eurocode 8 [14] and those for the deflection (w) under the horizontal force at the top of the wall element are obtained by using the described procedure on the samples (Fig. 3.44). Further details and analysis of the numerical results can be found in Kozem Šilih [12].

As mentioned above, a large proportion of walls (over 50 %, or even up to 80 %) may consist of elements with openings. A simple calculation based on the results presented in Fig. 3.44 shows that in a wall without any openings (G2)— 50 % and those with openings (O1)—50 %, approximately 13 % of the entire horizontal load would be transmitted to O1 elements. In the case of a wall with 80 % of elements O2 (and 20 % of G2), the ratio of the force resisted by elements O2 rises to 54 %. Elements with openings may transmit a considerable share of the horizontal load, depending on the number and types of different elements in the wall system.

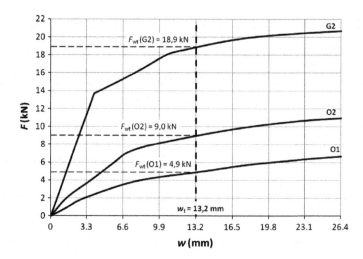

Fig. 3.44 F-w diagrams with the values of forces at the target displacement w_t (0, 5 % of the storey height)

The following are highly significant conclusions drawn about the influence the openings exert on structural behaviour of timber-frame buildings under the horizontal load:

- In the initial phase, the structure is elastic. Additional stiffness provided by the wall elements with openings results in a decrease in the load to which the solid elements are exposed and the first cracking occurs at a higher value of the horizontal load. The elastic resistance of the structure (until the point of the formation of irreversible damage) is therefore improved.
- After the formation of the cracks in the full wall elements, the elements with openings remain elastic for an additional amount of time, which leads to the ductility of the structure being improved. Additional stiffness provided by the wall elements with openings in the post-elastic phase also results in a higher value of the ultimate load. Consequently, both the ultimate resistance (as a means of preventing structural collapse) as well as the so-called over strength capacity of the wall system are increased.

The above conclusions are of utmost importance and will serve as starting points in our further study in Sect. 4.4.

3.3.2 Two-Dimensional-Braced Frame Models

Timber-frame wall elements can be modelled as a sophisticated and most accurate finite element model—as a shell element model, or as a braced frame with one or two diagonals. In the past, two-dimensional (2D) braced elements were already used by different authors. Timber-frame wall modelling in prefabricated timber structures was presented by Kessel and zur Kammer [15] who introduced a truss model with two diagonals to simulate every wall panel. Horizontal stiffness of the wall was defined by the stiffness of the diagonals whose choice satisfied the following condition: the horizontal displacement at the top of the panel due to a horizontal action of 8 kN was $H/500$, where H represented the height of the wall element. zur Kammer [16] offers five different models to simulate timber-frame walls with a special computational program. The problem of modelling timber-frame walls was also discussed in Kessel and Schönhoff [17] who used a braced frame with a single fictive diagonal. The cross section and the modulus of elasticity of the diagonal were determined based on the assumption that deformation of the alternative system is equal to that of the shear field of the sheathing board. The shear stiffness of the shear field was calculated numerically, with the fasteners' distance being taken into account.

But none of the presented mathematical models for timber-frame walls includes both the bending and shear deformability of the timber-frame wall simultaneously. Such approach can be found in Pintarič and Premrov [18], where the problem of modelling is avoided with a braced frame model having one fictive diagonal (Fig. 3.45). Its effective cross section approximates the bending and the shear stiffness of the sheathing board as well as the flexibility of the fasteners. A braced frame acts as the cantilever beam with two rigid supports (Fig. 3.45). The cross section of the fictive diagonal is determined by the bending and shear stiffness of the timber-frame wall calculated with respect to Eurocode 5 [6], where the effective bending stiffness $(EI_y)_{eff}$ is determined by the »γ-method«, presented in Eq. (3.14). The flexibility of the fasteners is calculated through their stiffness coefficient γ_y, determined for mechanically joint beams, based on Eurocode 5 [6].

We proceed from the assumption that the horizontal displacement u_H of the existent timber-frame wall is equal to that of the model with the alternative diagonal, when the same horizontal force F_H is acting on both systems.

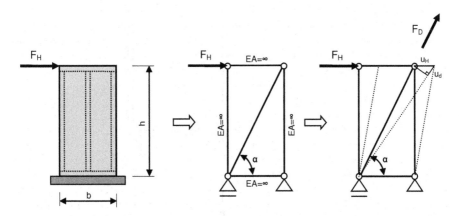

Fig. 3.45 Mathematical model of the timber-frame wall with one fictive diagonal

We furthermore presume that the axial stiffness of the surrounding frame is high enough to eliminate the impact of the frame, which would leave the flexibility of fictive diagonal as the only element to be taken into account.

Any material or cross section can be chosen for the fictive diagonal. The following is an example of a diagonal with a circular cross section, where the axial stiffness k_d of the fictive diagonal is determined as follows:

$$k_d = \frac{E_D \cdot A_{d,\text{fic}}}{L_D} \tag{3.6}$$

with E_D being the modulus of elasticity of the diagonal, $A_{d,\text{fic}}$ the fictive cross section of the diagonal and L_d the length of the diagonal. The connection between the stiffness of the existent timber-frame wall (k_p) and the stiffness of the fictive diagonal (k_d) is derived according to Fig. 3.46 and with the prediction of small displacements u_H. A final expression for the relation between k_p and k_d is given in the form of

$$k_p = \frac{F_H}{u_H} = \frac{F_d \cdot \cos\alpha}{\frac{u_d}{\cos\alpha}} = \frac{F_d}{u_d} \cdot \cos^2\alpha = k_d \cdot \cos^2\alpha \tag{3.7}$$

Combining Eqs. (3.6) and (3.7) gives the expression for the cross section of the fictive diagonal $A_{d,\text{fic}}$:

$$A_{d,\text{fic}} = \frac{k_p \cdot L_d}{E_D \cdot \cos^2\alpha} \tag{3.8}$$

Fig. 3.46 Scheme of the force distribution in a timber-frame wall element

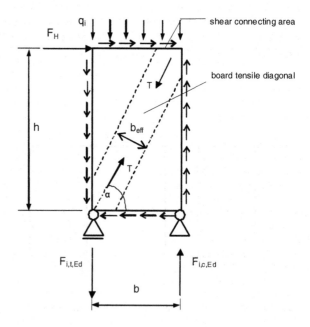

The diameter d_{fic} of the fictive diagonal of the circular cross section is then defined as follows:

$$d_{\text{fic}} = 2 \cdot \sqrt{\frac{A_{d,\text{fic}}}{\pi}} \qquad (3.9)$$

Based on the assumption of equal horizontal displacements of the existent timber-frame wall and the model with the fictive diagonal, we can claim that the axial stiffness of the diagonal is directly connected with the stiffness of the timber-frame wall obtained by the analytical calculation. As seen in Eq. (3.9), the cross section of the fictive diagonal $A_{d,\text{fic}}$ includes and simulates random staple distance s_{eff}, cracks of the sheathing boards (with EI_{eff} being reduced) as well as the openings and fixed glazing (with the reduced stiffness of the timber-frame wall). The height of the timber-frame wall, timber quality, the dimension of the studs and the type of the sheathing board can also be randomly selected.

3.3.2.1 Accuracy of the Model: Numerical Example

Horizontal displacements of the timber-frame wall at the horizontal force $F_H = 10$ kN, calculated through a mathematical model with the fictive diagonal, are compared with the already known displacements from previous FEM models described in Sect. 3.3.1. Table 3.10 features a comparison of the displacements for different staple distances (37.5, 75 and 150 mm) obtained by using the following procedures:

- The finite element method using the SAP 2000 program, taken from Kozem Šilih et al. [13]
- Experimental studies taken from Premrov and Kuhta [19].

The above table shows highly comparable values of horizontal displacements obtained through different methods. If the finite element model is taken as the most accurate, the deviation of the horizontal displacement in the braced frame with the fictive diagonal from that in the FEM model ranges from 0.9 % for the staple distances of 150 mm to 8 % for the distance of 37.5 mm. The difference is small in number, although rather significant in terms of a time-saving aspect and simplicity of the model.

Table 3.10 Values of horizontal displacements (mm) obtained through different methods

Horizontal displacement $F_H = 10$ kN (mm)	Braced frame with fictive diagonal	FEM model (SAP 2000)	Experimental results
$s_{\text{eff}} = 37.5$	2.35	2.55	2.65
$s_{\text{eff}} = 75$	2.86	2.92	2.75
$s_{\text{eff}} = 150$	3.42	3.45	4.55

To sum up, the mathematical model with the fictive diagonal offers a very good approximation in the calculation of the timber-frame wall element horizontal stiffness. The simplicity makes this model applicable also in combination with easily manageable and cheaper programs. In addition, a lot of time is saved in comparison with more complex and detailed modelling with the finite element method using more expensive and complicated programs, as presented in Sect. 3.3.1. When braced frames are combined, the entire system behaves as a structure made of separate independent cantilever beams modelled with the help of fictive diagonals. Braced frame modelling requires the knowledge of the bending and shear stiffness of the wall element which is determined with the analytical calculation, according to Eurocode 5 [6] and a semi-analytical procedure described in Premrov and Dobrila [20]. Walls with a different stiffness value are modelled with fictive diagonals of different thickness. Eventual appearance of the cracks in the fibre-plaster boards is simulated by reducing I_b and using the mathematical model from Premrov and Dobrila [20]. Another benefit of this method, vitally important for our study presented in Sect. 4.3, is its suitability for timber-frame wall elements with openings or fixed glazing, where the reduced stiffness of the wall is simulated with an adequately reduced diameter of the fictive diagonal.

3.3.3 Semi-Analytical Simplified Shear Models

In practice, there is often a need for simple and useful expressions to determine the racking resistance of the timber-frame walls by using hand calculation only, without recourse to computational programs. Two examples of such models will be presented in Sects. 3.3.3 and 3.3.4.

Källsner [21] and Äkerlund [22] suggested an interesting approach to determine the load-bearing capacity of the wall unit, based on two major assumptions:

- Behaviour of the joints between the sheet and the frame members is assumed to be linear elastic until the point of failure.
- The frame members and the sheets are assumed to be rigid and hinged to each other.

Furthermore, two simplified computational methods are given in Eurocode 5 [6] in order to determine the load-bearing capacity of the wall diaphragm. The first, *Method A*, is identical to the "Lower bound plastic method", presented by Källsner and Lam [23]. It defines the characteristic shear resistance of the wall ($F_{v,Rk}$) as a sum of all the fasteners' shear resistance values ($F_{f,Rk}$) along the loaded edges, based on the assumption that the timber-frame members and the sheets are rigid and hinged to each other:

$$F_{v,Rk} = \sum F_{f,Rk} \cdot \frac{b}{s} \cdot c \qquad (3.10a)$$

$$c = \begin{cases} 1 & \text{for } b \geq b_0 \\ \dfrac{b}{b_0} & \text{for } b \leq b_0 \end{cases} \qquad \text{where } b_0 = h/2 \qquad (3.10b)$$

The second simplified analysis, *Method B*, is applicable only to walls made from sheets of timber-based panel products fastened to the timber frame by means of nails or screws, with the fasteners being evenly spaced around the perimeter of the sheet. According to Method A, the sheathing material factor (k_n), the fastener spacing factor (k_s), the vertical load factor ($k_{i,q}$) and the dimension factors for the panel (k_d) are included in the design procedure in the form of:

$$F_{v,\mathrm{Rk}} = \sum F_{f,\mathrm{Rk}} \cdot \frac{b_i}{s_0} \cdot c_i \cdot k_d \cdot k_{i,q} \cdot k_s \cdot k_n \qquad (3.11a)$$

$$s_0 = \frac{9,700 \cdot d}{\rho_k} \qquad (3.11b)$$

where d is the diameter of the fastener and ρ_k the characteristic density of the timber frame. A scheme of the force distribution in a timber-frame wall unit under the horizontal force acting at the top of the element is presented in Fig. 3.46.

Eurocode 5 [6] simplified methods are said to be suitable for any fastener spacing commonly used in production in the case of using timber-based sheathing boards. However, these methods are not applicable in cases where the fibre-plaster sheathing material is used and where the tensile strength is very low as the cracks usually appear before the stresses in the fasteners achieve the yielding point. Källsner and Girhammar [24] developed an elastic analysis model for fully anchored sheathed timber-frame shear walls with timber-based sheathing boards. The model is based on the assumption of a linear elastic load-slip relation for the sheathing-to-framing joints. To suit the requirements of a special commonly appearing case, where $h_i = 2b_i$ and $s = s_r = s_p = s_{is}/2$, the authors developed a simplified expression for the horizontal load-bearing capacity (F_H) in the form of

$$F_H \approx \frac{1}{6} \cdot \sqrt{\frac{5,929}{170}} \cdot \frac{b_i}{s_r} \cdot F_v \approx 0.984 \cdot \frac{b_i}{s} \cdot F_v \qquad (3.12)$$

where F_v denotes the shear capacity of the fastener. If the lateral stress on the fastener achieves the characteristic yield point, the expression approximates Eq. (3.11a) defined by Eurocode. The authors additionally studied the effect of flexible framing members, as well as the effect of shear deformation of the sheet and that of vertical loads.

3.3.4 Semi-Analytical Simplified Composite Beam Models

The treated wall elements consisting of the timber frame and the sheathing boards behave in reality as composite elements and should thus be analysed as such. With the (sheathing) board taking the function of a coating material, the horizontal load shifts part of the force via mechanical fasteners to the fibre-plaster boards and the wall acts as a deep composite beam. Distribution of the horizontal force in composite treatment of the element depends on the proportion of stiffness. The effective bending stiffness $(EI_y)_{\text{eff}}$ of mechanically jointed beams, which empirically considers the flexibility of fasteners via coefficient γ_y—taken from Eurocode 5 [6], can be written in the form of

$$(EI_y)_{\text{eff}} = \sum_{i=1}^{n} E_i \cdot \left(I_{yi} + \gamma_{yi} \cdot A_i \cdot a_i^2\right) = \sum_{i=1}^{n_{\text{timber}}} \left(E_i \cdot I_{yi} + E_i \cdot \gamma_{yi} \cdot A_i \cdot a_i^2\right)_{\text{timber}}$$

$$+ \sum_{j=1}^{n_b} \left(E_i \cdot I_{yi}\right)_b \tag{3.13}$$

where n is the total number of elements in the relevant cross section, and a_i is the distance between the global y-axis of the entire cross section and the local y_i-axis of the ith element with a cross section A_i (see Fig. 3.47). The equation shows that the bending stiffness $(EI_y)_{\text{eff}}$ strongly depends on the stiffness coefficient of the fasteners (γ_y). With respect to Eurocode 5, γ_y can be defined via the fastener spacing (s) and the slip modulus per shear plane per fastener (K) in the form of

$$\gamma_y = \frac{1}{1 + \left(\frac{\pi^2 \cdot A_{t1} \cdot E_t \cdot s}{L_{\text{eff}}^2 \cdot K}\right)} \tag{3.14}$$

The expression of the so-called "γ-method" is based on the differential equation for the partial composite action with the following fundamental assumptions:

- Bernoulli's hypothesis is valid for each subcomponent.
- The slip stiffness is constant along the whole connecting area in the element.
- Material behaviour of all subcomponents is considered to be linear elastic (Fig. 3.47).

The simplified solutions are closely related to Möhler's model [25] which takes into consideration the fact that a beam is simply supported and has constant slip stiffness. Four basic assumptions have been additionally considered through modelling of the cracked cross section in the sheathing boards of the composite beam model, Premrov and Dobrila [20]:

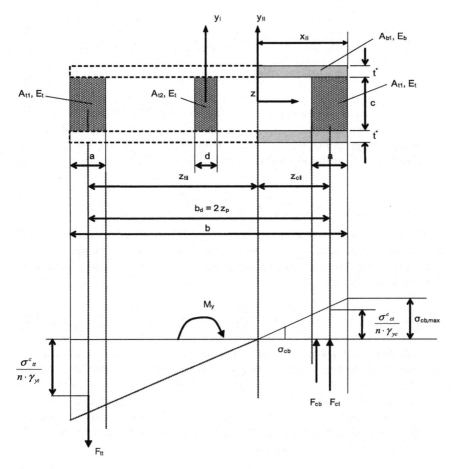

Fig. 3.47 Mathematical model and the normal stress distribution in the cross section

- The tensile area of the boards is neglected after the first crack formation.
- Linear elastic material behaviour of the compressed area of the boards and the timber frame is assumed.
- The stiffness coefficient of the fasteners in the tensile connecting area ($\gamma_{yt,mod}$) is presumed to remain constant after the appearance of the first crack.
- The stiffness coefficient of the fasteners in the compressed connecting area ($\gamma_{yc,mod}$) is not constant and depends on the lateral force acting on a single fastener.

An approximate semi-analytical mathematical model is schematically presented in Fig. 3.47. Using the classical beam theory the characteristic horizontal force forming the first tensile crack ($F_{cr,k}$) in the sheathing board is defined according to the normal stress criteria and the characteristic tensile strength of the board ($f_{bt,k}$):

$$\sigma_{b,max} = \frac{M_y \cdot E_b}{(EI_y)_{eff}} \cdot \frac{b}{2} \quad \Rightarrow \quad M_{y,cr,k} = \frac{2 \cdot f_{bt,k} \cdot (EI_y)_{eff}}{E_b \cdot b} \tag{3.15a}$$

$$F_{cr,k} = \frac{M_{y,cr,k}}{h_d} = \frac{2 \cdot f_{bt,k} \cdot (EI_y)_{eff}}{E_b \cdot b \cdot h_d} \tag{3.15b}$$

If we declare a characteristic destruction condition of the wall element the failure case when the tensile stress in timber ($\sigma_{tt,max}$) achieves the characteristic tensile timber strength ($f_{t,0,k}$), the characteristic horizontal destruction force ($F_{u,k}$) is computed in the following form:

$$F_{u,k} = \frac{f_{t,0,k} \cdot (EI_y)_{eff}^{II}}{E_t \cdot \left(\gamma_{yt} \cdot z_{tII} + \frac{a}{2}\right) \cdot h} \tag{3.16}$$

where $(EI_y)_{eff}^{II}$ represents the effective bending stiffness of the cross section with a crack in the tensile section of the board (Fig. 3.47). Thus, the lateral force (F_1) acting on a single fastener in a single shear can be calculated in dependence on the horizontal point load (F_H):

$$F_1 = \frac{(ES_y)_{eff}}{(EI_y)_{eff}} \cdot \frac{s}{2} \cdot F_H \tag{3.17}$$

As designers and producers usually find themselves in an important dilemma of deciding upon the best sheathing board (OSB or FPB) to be used in a building according to its height and location, our further attention will be focused mainly on the analysis of the sheathing boards' influence on the racking resistance of the wall elements. The results will serve as confirmation of the applicability of both simplified models presented. A detailed numerical study of the above issue can be found in Premrov and Dobrila [26].

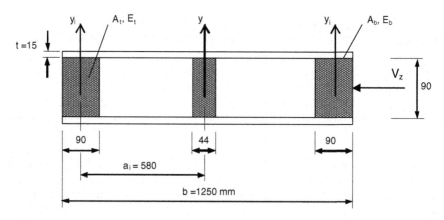

Fig. 3.48 Cross section of the test sample

Table 3.11 Material properties of the materials used

	$E_{0,m}$ (N/mm²)	G_m (N/mm²)	$f_{m,k}$ (N/mm²)	$f_{t,0,k}$ (N/mm²)	$f_{c,0,k}$ (N/mm²)	$f_{v,k}$ (N/mm²)	ρ_m (kg/m³)
Timber C22	10,000	630	22	13	20	2.4	410
FPB	3,000	1,200	4.0	2.5	20	5.0	1,050
OSB 3	3,500	240	20	20	20	/	600

Numerical analyses are performed on the wall element with dimensions of $h_i = 2{,}635$ mm and $b_i = 1{,}250$ mm, consisting of timber studs ($20 \times 90 \times 90$ mm and $10 \times 44 \times 90$ mm) and timber girders ($20 \times 80 \times 90$ mm). The sheathing boards (OSB or FPB) with a thickness of $t = 15$ mm are fixed to the timber frame by means of staples with a diameter of $d = 1.53$ mm and a length of $l = 35$ mm.

Fig. 3.49 Characteristic racking resistance of the wall in dependence of the fasteners' spacing (s) (**a**) DSB (**b**) FPB

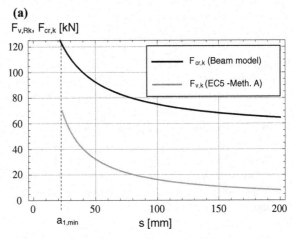

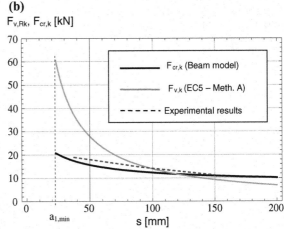

The composite cross section of the element is schematically presented in Fig. 3.48. Eurocode 5 [6] sets the minimum spacing for staples at $a_{1,\min} = (10 + 5|cos\,\alpha|)d = 15d = 22.95$ mm. Our analysis will therefore be presented with respect to the above range.

Material properties of timber with a strength grade C22 are taken from EN 338:2009, FPB material properties originate from Knauf [27] and those of type 3-OSB from Kronopol [28] and EN 300:2006 [36]. All the values are listed in Table 3.11.

Both calculated characteristic racking resistances in dependence of the fastener spacing (s), obtained through Eq. (3.10a) for the wall element with OSB sheathing boards and through Eq. (3.15a) for the wall element with FPB sheathing boards, are graphically presented in Fig. 3.49a and b. The values measured for FPB are taken from Premrov and Kuhta [29]; they were obtained with test samples having the same material characteristics and dimensions as those in the numerical study.

As seen from Fig. 3.49a, the value $F_{v,\mathrm{Rk}}$ in the case of OSB sheathing boards is far under the force causing the first crack in the board ($F_{\mathrm{cr},k}$) at any value of the fastener's spacing (s). The reason lies in the fact that the tensile strength of OSB boards is relatively high and equal to the compressive strength (see Table 3.11). According to Eq. (3.15b), the value obtained for $F_{\mathrm{cr},k}$ is unrealistically high and should not be considered. Consequently, the fastener yielding in the wall elements with OSB boards is decisive for almost any value of s, which calls for Eq. (3.10a) to be used. The difference between the values for $F_{\mathrm{cr},k}$ and $F_{v,\mathrm{Rk}}$ strongly depends on the fastener's spacing and proves to be smaller at a lower value of s.

In FPB sheathing boards (Fig. 3.49b), there is evidence of both methods being in relatively good agreement only at the fastener spacing ranging from 100 to 150 mm. However, it needs to be underlined that such spacing is usually not used in production. The case of FPB sheathing boards demands a particular respect of the values obtained through the composite beam model on account of their better comparability with the measured values.

Fig. 3.50 Racking resistance of walls with FPB and OSB sheathing boards

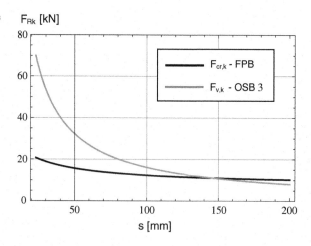

To sum up the current subsection, it would be interesting to compare the racking resistance with the values obtained by expressions which are decisive for each type of sheathing boards. According to $F_{v,k}$, the shear model is decisive for the walls with OSB sheathing, while the composite beam model has a decisive role in FPB sheathing boards, according to $F_{cr,k}$. The functions of F_{Rk} in dependence of s are presented in Fig. 3.50.

It is obvious that the racking resistance for FPB is far under the values for OSB. The difference increases with a smaller fastener spacing. In addition, fastener spacing has a stronger influence on the wall elements with OSB sheathing boards.

3.4   Multi-Storey Timber-Frame Building

There is a worldwide increasing tendency to build multi-storey (three and more) prefabricated timber-frame buildings with timber-frame wall panels functioning as the main bearing capacity elements. A Slovenian example is a six-storey Hotel Dobrava on the Pohorje Mountain, built in 2000 (Fig. 3.51), with overall ground plan dimensions of 38 × 18.60 m, Premrov [30]. The first two storeys, of which one is underground, were built in a classical system with thin concrete walls. The rest of the buildings, the upper four storeys, were erected in the timber-frame-panel wall system, thoroughly described in Sect. 3.2.3.4.

Seismic and windy areas witness an essential increase in the horizontal load actions and a consequent direct, strong impact on the internal forces in the load-bearing timber-frame-panel wall elements. For a wall assembly placed in one principal direction, consisting of one or more wall elements, a simplified static design presented in Fig. 3.52 can be used to calculate the axial forces (N), the shear forces (V) and the bending moments (M) due to vertical and horizontal loads.

The force acting at the top of a single wall element is obtained by using the procedure presented in Sect. 3.3, Fig. 3.40. As mentioned in Sect. 3.3.4, the timber-frame-panel walls with fibre-plaster sheathing boards (FPB) can be treated

Fig. 3.51 A six-storey Hotel Dobrava in Terme Zreče, the Pohorje Mountain, Slovenia

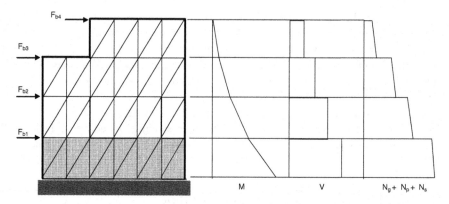

Fig. 3.52 Static design for the horizontal load action in one principal direction for the frame-panel building of Hotel Dobrava

as composite elements. Distribution of the horizontal force through composite treatment of the element depends on the proportion of the stiffness. As the tensile strength of the FPB is approximately 10 times lower than the compressive strength, and evidently smaller than timber strength of all members in the timber frame, the FPB usually act as a weaker part of the presented composite system. Thus, multi-storey buildings located in seismic or windy areas usually face the appearance of cracks in the FPB. The latter consequently lose their stiffness and their contribution to the total horizontal stiffness of the whole wall assembly should therefore not be taken into account. Stresses in the timber frame under the horizontal loads are usually not critical. In such cases, it is necessary to strengthen the wall elements to assure the horizontal stability of the structure. There are several possibilities of reinforcing arrangements:

- Using additional boards which are usually doubled:

  – Symmetrically (on both sides of the timber frame)
  – Non-symmetrically (on one side of the timber frame)

- By reinforcing boards with steel diagonals
- By reinforcing boards with carbon or high-strength synthetic fibres (FRP, CFRP, etc.).

In Dobrila and Premrov [31], experimental results using *additional fibre-plaster boards (FPB)* are presented. The test samples demonstrated higher elasticity, while the bearing capacity and ductility, in particular, were not improved in the desired range.

With the intention of improving the resistance and, especially, the ductility of the walls, it is more convenient to insert classical *diagonal steel strips* which have to be fixed to the timber frame. A part of the horizontal force is thus shifted from the boards via the tensile steel diagonal to the timber frame, after the appearance of the first crack in the tensile zone of the FPB. Since the strips need to be connected

Table 3.12 Measured experimental results

Type of the test samples	F_{cr} (kN)	$F_{u,k}$ (kN)	$q = F_{u,k}/F_{cr}$	$q_d = u_{u,k}/u_{cr}$
FPB ($s = 150$ mm)	11.27	16.94	1.50	2.50
FPB ($s = 91$ mm)	14.65	20.12	1.37	1.80
FPB ($s = 37.5$ mm)	18.94	48.42	2.56	3.61
FPB ($s = 75$ mm)	17.06	26.17	1.53	2.71
FPB—double boards ($s = 75$ mm)	23.98	32.67	1.36	1.65
FPB + BMF steel strips	18.50	35.73	1.93	2.97
FPB + CFRP 300 mm glued to the timber	24.28	40.33	1.66	2.80
FPB + CFRP 300 mm	35.90	36.26	1.01	2.71
OSB 3 ($s = 75$ mm)	$F_y = 21.25$	41.55	1.96	4.19
Type of the opening				
FPB—window	7.16	12.85	1.80	3.86
57.2 × 127.2 cm				
FPB—window	4.30	8.91	2.07	5.03
84.2 × 127.2 cm				

FPB—fibre-plaster board
OSB—oriented strain board
S—distance between fasteners in the connecting plane between timber frame and FPB
F_{cr}—force forming the first crack in the sheathing board
F_y—force forming the fasteners yielding
$F_{u,k}$—ultimate failure force

to the timber-frame elements, a special hole has to be made in each timber corner in order to place the diagonals and the boards in the same plane. The strips are then nail-fastened to the timber elements, as in the case of the reinforcement arrangement at Hotel Dobrava. The first storey of the timber part of the building was reinforced within this concept to assure the horizontal stability of the entire building, Premrov [30]. Although the type of reinforcement described is technologically rather time-consuming, it tops the list of available strengthening solutions to gain the best ductility (Table 3.12). The data from the table also clearly point to the fact that the measured forces forming the first crack (F_{cr}) in the FPB are not essentially higher in comparison with the non-reinforced test samples. There is furthermore hardly any influence on the element stiffness produced by any type of reinforcement prior to the appearance of cracks in the non-strengthened FPB. Nevertheless, the ratio between the measured failure forces (F_u) shows that the resistance and ductility of the reinforced panels increase by 77 and 39 %, respectively, which is essential for the seismic behaviour of the elements.

Owing to the time-consuming nature of the technological concept involving classical steel diagonal strips, we tried to find another, more practical solution to strengthen the fibre-plaster boards (FPB)—using *CFRP strips* which are glued to the FPB in the tensile diagonal direction (Fig. 3.53a, b). This strengthening concept is applied in order to make CFRP composites contribute to the tensile capacity when the tensile strength of the FPB is exceeded. Experimental results presented in Premrov et al. [32] demonstrated several important facts. Similarly to the steel diagonal concept, there was hardly any influence on the element stiffness produced

Fig. 3.53 Strengthening concept by gluing CFRP strips in the diagonal direction to the FPB (**a**), the strips are not glued to the timber-frame elements (**b**)

by any type of reinforcement prior to the appearance of cracks in the non-strengthened FPB. Nonetheless, after the first cracks in the non-strengthened FPB appeared, the test samples demonstrated an important difference in behaviour dependant on the boundary conditions between the inserted CFRP strips and the timber frame. When the strips were glued to the boards and additionally to the timber frame (Fig. 3.53a), the fasteners produced a substantially smaller slip in the connecting area, which never exceeded a distance of 1 mm when the first tensile cracks in the FPB appeared. On the other hand, in the case where the CFRP diagonals were not fastened to the timber frame (Fig. 3.53b), the slip between the FPB and the timber frame was evidently larger and the walls tended to fail because of the fastener yielding.

Another point of significance is seen in the contribution of CFRP diagonal strip reinforcement to the load-bearing capacity which proved to have a relatively high value (cf. the results measured—Table 3.12). The current high costs of applying CFRP are justified by the experimental results proving that significantly higher forces forming the first crack lead to the load-bearing capacity and stiffness being increased.

To have a better insight into the presented strengthening concepts, Table 3.12 shows all the measured results taken from our previous experimental studies, Dobrila and Premrov [31], Premrov et al. [32], Premrov and Kuhta [19]. The results obtained via test samples with openings (Fig. 3.43) by Kozem Šilih and Premrov [33] are additionally included with the intention to provide a better understanding of the influence of the openings on the horizontal stability and ductility of the entire building. Further specifics referring to the above studies are available in the listed references.

A practical point of view based on the presented results is seen in the fact that the wall elements with OSB sheathing boards have a generally higher racking resistance than those with FPB. Therefore, it can be recommended to use OSB sheathing boards in the case of tall timber-framed buildings located in heavy windy or seismic areas.

Moreover, it is obvious that the wall elements with openings demonstrate an essential decrease in the load-carrying capacity. Consequently, stability problems may appear in tall timber buildings with an enlarged size of door or window openings, especially at lower levels of the building where a more powerful impact of the horizontal loads can occur (Fig. 3.52). A special analysis is needed in such cases and the wall elements with FPB boards have to be adequately strengthened with the presented concepts of reinforcement to improve the stability of the whole structure.

Finally, there is a problem of assuring the horizontal stability of the structure having an enlarged size of fixed glazing in addition to doors and windows, especially when these need to be in the lower part of the building. This issue will be dealt with in Chap. 4.

References

1. Gold S, Rubik F (2009) Consumer attitudes towards timber as a construction material and towards timber frame houses—selected findings of a representative survey among the German population. J Cleaner Prod 17:303–309
2. Lokaj A (2007) Montonovane drevostavby—Moderni zpusob bydleni, Rozvoj dřevěných konstrukcí. In: Sborník příspěvků z odborného semináře, Brno, 19 Apr 2007
3. Kolb J (2008) Systems in timber engineering. Birkhäuser Verlag AG, Basel—Boston—Berlin
4. Deplazes A (2005) Constructing architecture—materials process structures, a handbook. Birkhäuser—Publishers for Architecture, Basel, Switzerland
5. Augustin M (2008) Wood based panels. In: Handbook 1—timber structures, Leonardo da Vinci Pilot project CZ/06/B/F/PP/168007, Educational materials for designing and testing of timber structures
6. European Committee for Standardization CEN/TC 250/SC5 N173 (2005) EN 1995-1-1:2005 Eurocode 5: design of timber structures, Part 1-1 general rules and rules for buildings, Brussels
7. Deplazes A, Fischer J, Ragonesi M (2007) Lignatec Massivholzbau—Die technischen Holzinformationen der Lignum. Lignum, Holzwirtschaft Schweiz, Zürich
8. Breyer ED, Fridley JK, Cobeen EK, Pollock GD (2007) Design of wood structures—ASD/LRFD, 6th edn. McGraw-Hill Publishing Company
9. Faherty KF, Williamson TG (1989) Wood engineering and construction handbook. Mc Graw-Hill Publishing Company
10. Thelandersson S, Larsen JS (2003) Timber engineering. Wiley, London
11. Schulze H (2005) Wände—Decken—Bauproducte—Dächer—Konstruktionen—Bauphysik—Holzschutz. 3. Auflage. B.G. Teuber Verlag
12. Kozem Šilih E (2012) PhD thesis
13. Kozem Šilih E, Premrov M, Šilih S (2012) Numerical analysis of timber-framed wall elements coated with single fibre-plaster boards. Eng Struct 41:118–125
14. European Committee for Standardization CEN/TC 250 (2005) EN 1998-1:2005 Eurocode 8: design of structures for earthquake resistance, Part 1 general rules, seismic actions and rules for buildings, Brussels
15. Kessel HM, zur Kammer T (2004) Three-dimensional load-bearing behaviour of multi-storey timber frame buildings. In: Proceedings of the 8th world conference on timber engineering, vol 1, pp 7–12, Lahti Finland

16. zur Kammer T (2006) Zum räumlichen Tragverhalten mehrgeschossiger Gebäude in Holztafelbauart. Dissertation. Institut für Baukonstruktion und Holzbau der Technischen Universität Carolo-Wilhelmina zu Braunschweig; Braunschweig

17. Kessel M, Schönhoff T (2001) Entwicklung eines Nachweisverfahrens für Scheiben auf der Grundlage von Eurocode 5 und DIN 1052 neu. Forschungsbericht des Instituts für Baukonstruktion und Holzbau. Braunschweig

18. Pintarič K, Premrov M (2012) Mathematical modelling of timber-framed walls using fictive diagonal elements. Submitted for publication in applied mathematical modelling

19. Premrov M, Kuhta M (2009) Influence of fasteners disposition on behavior of timber-framed walls with single fibre-plaster sheathing boards. Constr Build Mater 23(7):2688–2693

20. Premrov M, Dobrila P (2008) Mathematical modelling of timber-framed walls strengthened with CFRP strips. Appl Math Model 32(5):725–737

21. Källsner B (1984) Panels as wind-bracing elements in timber-framed walls. Swedish Institute for Wood Technology Research, Report 56, Stockholm

22. Äkerlund S (1984) Enkel beräkningsmodell för skivor på regelstomme (Simple calculation model for sheets on a timber frame). In: Bygg and Teknik, No.1

23. Källsner B, Lam F (1995) Diaphragms and shear walls. Holzbauwerke: Grundlagen, Entwicklungen, Ergänzungen nach Eurocode 5, Step 3, Fachverlag Holz, Düsseldorf, p.15/1-15/19

24. Källsner B, Girhammar UA (2009) Analysis of fully anchored light-frame timber shear walls-elastic model. Mater Struct 42(3):301–320

25. Möhler K (1956) Über das Tragverhalten von Biegeträgern und Druckstäben mit zusammengesetztem Querrshnitt und nachgiebigen Verbindungsmitteln. Habilitation, TH Karlsruhe

26. Premrov M, Dobrila P (2011) Numerical analysis of sheathing boards influence on racking resistance of timber-frame walls. Adv Eng Softw 45(1):21–27

27. Knauf (2002) Gipsfaserplatten Vidiwall/Vidifloor, Knauf LLC, Dubai, United Arab Emirates

28. Kronopol Swiss Krono Group, Kronotec AG (2004) Kronopol OSB, Luzern

29. Premrov M, Kuhta M (2010) Experimental analysis on behaviour of timber-framed walls with different types of sheathing boards. Construction materials and engineering. Nova Science Publishers, New York

30. Premrov M (2008) Sport hall Rogla, case study no. 13. In: Educational materials for designing and testing of timber structures—TEMTIS, case studies, instruction handbook. Ostrava: VŠB-TU, Fakulta stavební, http://fast10.vsb.cz/temtis/documents/Instruction_13_Rogla.pdf

31. Dobrila P, Premrov (2003) Reinforcing methods for composite timber frame-fiberboard wall panels. Eng Struct 25(11):1369–1376

32. Premrov M, Dobrila P, Bedenik BS (2004) Analysis of timber framed walls coated with CFRP strips strengthened fibre-plaster boards. Int J Solids Struct 41(24/25):7035–7048

33. Kozem Šilih E, Premrov M (2010) Analysis of timber-framed wall elements with openings. Constr Build Mater 24(9):1656–1663

34. European Committee for Standardization CEN/TC 250 (2004) EN 1992-1-1 Eurocode 2: design of concrete structures, Part 1-1 general rules and rules for buildings, Brussels

35. European Committee for Standardization (2009) EN 338:2003 E: structural timber—strength classes, Brussels

36. European Committee for Standardization (2006) EN 300:2006

Chapter 4
Timber-Glass Prefabricated Buildings

Abstract This chapter is based on using timber and glass which were formerly rather neglected as construction materials. With suitable technological development and appropriate use, they are nowadays becoming essential construction materials as far as energy efficiency is concerned. Their combined use is extremely complicated, from the energy efficiency perspective presented in Sect. 4.3 on the one hand and from the structural viewpoint presented in Sect. 4.4 on the other, which sets multiple traps for designers. A good knowledge of their advantages and drawbacks is thus vitally important, as will be seen in the first two sections of the current chapter. The results of the comparative analysis contained in the last two sections can serve as a good frame of reference to architects and civil engineers in their approximate estimation of the energy demands emerging from different positions and proportions of the glazing surfaces, in addition to being of assistance in their assessment of the influence of the building shape exerts on the energy demand of prefabricated timber-frame buildings.

4.1 The History of Glass Use

The origins of glass manufacture are still uncertain. Historical sources and archaeological evidence trace them to Mesopotamia or Ancient Egypt. In general, the first glass-made objects (where glass was used as an independent material) appeared no sooner than around 3,000 BC [1] and were mainly meant as decoration. Manufacturing glass vessels and other useful items began only much later.

In the second century BC, the Phoenicians made a revolutionary discovery of glass blowing which permitted limitless ways of glass-shaping. A major turning point in the development of glassmaking goes back to the times of the Roman Empire and the Romans who were the first manufacturers of clear glass. Glass-blowing as a glass-forming technique of Syrian origin replaced the old Roman glassmaking techniques.

V. Žegarac Leskovar and M. Premrov, *Energy-Efficient Timber-Glass Houses*, Green Energy and Technology, DOI: 10.1007/978-1-4471-5511-9_4, © Springer-Verlag London 2013

The Romans began using glass for architectural purposes and introduced the first instances of the glazing in the exterior wall openings. The spreading of the glassmaking technique reached also northern Alpine regions and after the collapse of the Western Roman Empire it continued to develop in the near east and the world of Islam. The seventh century witnessed a discovery of a simple and inexpensive window glassmaking method, whereas the eighth century saw the appearance of coloured glass which was later used for stained glass windows in churches. From the fifteenth to the seventeenth century, glassmaking continued its development in the Venetian Republic which remained a glass-manufacturing centre until the eighteenth century. In the sixteenth century, the production of glass stretched further over European countries among which Bohemia distinguished itself by making a most intense headway in the field of glass manufacturing. Two of the most important hand-blown window glass production techniques dating from the early Middle Ages included the cylinder blown sheet and the crown glass. Both remained basic glass-manufacturing methods until the late nineteenth and the early twentieth centuries [2].

The flourishing of the glass-manufacturing industry was marked by technological advances. Innovative procedures of glassmaking kept emerging since 1687 when a Frenchman, Perrot introduced the cast glass process which enabled the production of glass panes in the size of 1.2×2.0 m. Glass became an item of mass production only towards the end of the Industrial Revolution. In 1839, the Chance brothers improved the process of making cylinder blown sheet glass. Further innovations were Siemens' patent of the melting furnace in 1856, followed by Lubbers' development of the mechanical process to combine blowing and drawing, appearing at the end of the nineteenth century, and by Owen's invention of the machine for blowing of bottles, dating from the same period. A key person and a father of modern research of glass was F.O. Schott, a German chemist and scientist whose scientific work conducted in the late nineteenth century involved methods for studying the effects of various chemical elements have on optical and thermal properties of glass. Other important improvements in glass production that followed in the twentieth century were the 1913 Fourcault process of drawing glass panes directly from the molten glass and the Libbey-Owens process, a similar method developed by Colburn, which enabled production of 2.5-metre-long glass panes of different thicknesses depending on the speed of the drawing. Combining both of the methods resulted in a process applied by the Pittsburgh Plate Glass Company in 1928, which led to an increase in the manufacturing speed. In 1919, the improved cast glass process invented by Bicheroux allowed the production of panes in the size of 3.0×6.0 m. However, one of the most important innovations was the 1950' Pilkington float-glass process which has been the basis of glass production until the present times [2].

Another milestone in the beginnings of modern glass construction was set by the development of skeletal iron construction. Both provided a technical basis for the erection of the first glazed load-bearing structures in the early nineteenth century, known as the English greenhouses. Stability of such delicate free-standing enclosures with domed and folded glazed roofs was achieved mainly with the bracing provided by small glass shingles embedded in the putty [3]. The progress in glass technology, which allowed the production of larger glass panes, led to the use of glass for covering purposes while its construction-related significance was almost completely lost. Throughout the twentieth century, the visual potential of glass was strongly emphasized through the phenomenon of transparent architecture focusing on large glass façades. The latter had a positive impact on daylight conditions and the aesthetic perception of internal spaces open to the exterior, while at the same time, the indoor living climate in buildings with large glazing areas was not a matter of high thermal comfort any more. Single-pane glazing used at that time caused large heat losses in winter in addition to unpleasant summer period heat gains resulting in overheating. A dynamic evolution of the glazing in the last 40 years has resulted in insulating glass products with highly improved physical features suitable for application in contemporary energy-efficient buildings.

An ongoing driving force of the development leads to the conclusion that glass as an important building material will not cease to be used in the future owing to its remarkable properties and multiple options of usage which have not yet been surpassed by any other material.

4.2 Glass as a Building Material

Glass can nowadays be treated as a dominating material in modern architecture. Not only its optical appearance but also its numerous technical functions classify glass as an indispensable building design component providing protection against atmospheric influences, natural lighting, energy storage and generation, etc. The energy aspect of this building material was often treated as a weak point in the past, whereas in the last few decades, it has become one of the main drivers of technical development of glass for the purposes of the building industry. Another important novelty arising from the phenomenon of increasing the size of glazing surfaces in modern buildings is taking the load-bearing function of glass into account, where its integration with other building materials plays a vital role.

Glass is an amorphous form of matter. Although different explanations about whether glass should be classified as a solid or as a liquid might have appeared over a time, it is important to define glass as a solid material, i.e., a solidified liquid.

The idea that glass is a liquid came from observing window glass in old churches. The glass being thicker at the bottom than at the top gives an impression as if the gravity had caused the glass to flow towards the bottom over several centuries, which in fact is not true. The difference in thickness arises from an ancient technique of the crown glass process, where glass sheets were thicker towards the edges.

Table 4.1 Chemical composition of predominantly used glass types

Substances	Chemical symbol	Share of substance in glass (%)
Silicon oxide	SiO_2	69–74
Calcium oxide	CaO	5–12
Sodium oxide	Na_2O	12–16
Magnesium oxide	MgO	0–6
Aluminium oxide	Al_2O_3	0–3

Glass is formed when the liquid is rapidly cooled from its molten state through its glass-transition temperature (T_g) into a solid state without crystallization [4]. Since the molecules of glass follow a completely random order and do not form a crystal lattice [5], its configuration is geometrically irregular, which gives glass its transparency. Upon heating, glass gradually changes from a solid to a plastic–viscous and finally to a liquid state. In comparison with timber, whose properties depend strongly on the direction of the grain, glass exhibits amorphous isotropy, i.e., its properties are uniform regardless of the direction of measurement.

Nowadays, the building industry predominantly uses soda-lime-silica glass (SLS). It consists of an irregular three-dimensional network where each silicon (Si) atom is bounded to four oxygen (O) atoms. The making of SLS involves four phases, i.e., preparation of raw material (soda ash, lime, silica sand and cullet), melting in a furnace, forming and finishing. Apart from float glass, which is generally used for windows, other products resulting from the above-described manufacturing process are container glass, pressed and blown glass. The composition of predominantly used glass types is presented in Table 4.1.

Apart from the basic substances, glass contains also small proportions of other substances, e.g., magnesium oxide and aluminium oxide, which provide additional influence on its colour and physical properties. Among the latter, the thermal conductivity (λ), the specific heat capacity (c), the transition temperature (T_g) as well as the average reflective index in the visible range of wavelengths (n) are of primary interest from the energy viewpoint of our further research. General physical properties of soda-lime-silica glass are presented in Table 4.2.

The thermal conductivity of soda-lime-silica glass is comparable to that of concrete whose value also ranges from 0.5 to 1.5 W/mK. The density and the coefficient of thermal expansion values of the two materials (c.f. Table 4.3) prove

Table 4.2 General physical properties of soda–lime–silica glass

Property	Symbol with units	Value
Transition temperature	T_g [°C]	564
Liquid temperature	T_l [°C]	1,000
Density	ρ [kg/m³]	2,500
Coefficient of thermal expansion	α_T [K^{-1}]	0.9×10^{-5}
Thermal conductivity	λ [W/(m K)]	1.0
Specific heat capacity	c [J/(kg K)]	720
Average reflective index in the visible range of wavelengths	n	1.5

Table 4.3 Mechanical characteristics of float glass, softwood and steel

	Density ρ [kg/m^3]	Compress strength f_c [N/mm^2]	Tensile bending strength f_{mt} [N/mm^2]	Modulus of elasticity E [N/mm^2]	Coefficient of thermal expansion α_T [10^{-5} K^{-1}]
Float glass	2,500.00	800.00	45.00	70,000.00	0.90
Timber C30	460.00	23.00	30.00	12,000.00	0.50
Steel S240	7,850.00	240.00	240.00	210,000.00	1.20
Concrete C30/37	2,500.00	30.00	2.9	33,000.00	1.00
Ratio glass/ timber	5.43	34.78	1.50	5.83	1.80
Ratio glass/ steel	0.32	3.33	0.18	0.33	0.75
Ratio glass/ concrete	1.00	26.67	15.52	2.12	0.90

to be a further similarity in addition to the relationship between the compressive and the tensile strengths (which will be further discussed in Sect. 4.2.1), with the compressive strength of both materials being very high and their tensile strength essentially lower. We can therefore presume that the behaviour of glass in many physical aspects demonstrates closely similar characteristics to those of concrete.

4.2.1 Structural Glass

Glass is a molecularly cooled liquid that the final stage of production turns into a solid. Owing to its optical and energy related, i.e., insulating properties, glass has become an ever more widely used material of the last decade and is no longer solely responsible for daylighting or the transparency of the building. Furthermore, its improved strength properties enabled the use of large glazing areas as additional structural resisting elements. Even though structural problems related to certain mechanical disadvantages of glass still exist, they can be generally avoided if glass elements are properly incorporated in the structural system of the building.

One of the main drawbacks of glass used as a load-bearing material lies in its being a relatively *brittle material* with mostly a significantly low degree of post-cracking resistance (Fig. 4.1a, b). When compared to the stress–strain diagram for timber in compression parallel to the grain (Fig. 3.8) and to that of timber in tension parallel to the grain (Fig. 3.11), glass demonstrates considerably lower ductility. As a consequence, resisting problems can occur in glass elements located in heavy seismic or windy areas. Post-cracking behaviour of glass depends on the type of glass, which is a matter of further discussion.

On the other hand, glass has a high *modulus of elasticity* of approximately 70 GPa, which is a value about 6 times higher than that of softwood in the grain

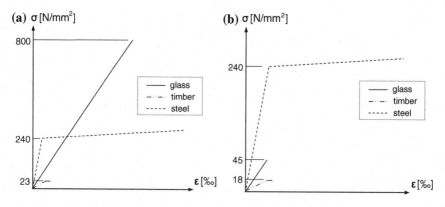

Fig. 4.1 σ–ε diagram of glass, timber and steel in compression (**a**), and tension (**b**)

direction, although 3 times lower than that of steel and equal to that of aluminium. Thus, we can claim that glass is relatively stiff material and that properly inserted glass elements can significantly contribute to the stiffness of the structure. A difficulty still remaining is seen in the behaviour of glass, which is almost linear-elastic until failure.

The *strength* of glass strongly depends on the type of loading. The strength of glass in compression is extremely high, mostly ranging from 700 to 900 MPa, which is about 2 times higher than in the case of steel and about of 40 times higher than the strength of timber. The tensile bending strength depends on the type of glass, but varies mostly from 45 MPa for float glass to 150 MPa for chemically strengthened glass. Both values prove to be far above those of the tensile strength of timber, but they are on the other hand far under those measured in steel.

Since most of the glass areas are placed in the south façade of the buildings and therefore subjected to a very high temperature effect, especially in the summer period, the *coefficient of thermal expansion* α_T is of vital importance, when glass and timber are used as composite elements. The values of α_T for glass, softwood and hardwood are 0.9×10^{-5}, 0.5×10^{-5} and 0.8×10^{-5} K^{-1}, respectively. Enlarged shear stresses may therefore occur in the adhesives between timber and glass elements when these are subjected to a heavy temperature effect.

Table 4.3 shows the most important mechanical values of float glass. They are compared to the values of softwood, concrete and steel.

It is obvious from the glass/timber, glass/steel and glass/concrete calculated ratios that two of the glass properties demonstrate substantial deviation from the density ratio of all four materials, namely the compressive strength of the glass which is extremely high and its tensile bending strength which is extremely low. The relationship of the values of the elasticity modulus is in good accordance with the relationship of the densities. The stress–strain diagrams (σ-ε) for glass, steel and timber are presented in Fig. 4.1a, b for the compression and tension, respectively. The linear simplification for timber is made according to the σ-ε diagram for compression in Fig. 3.8 and the σ-ε diagram for tension in Fig. 3.11.

There are many different types of structural glass:

- Float glass
- Annealed glass
- Heat-strengthened (partially tempered) glass
- Fully tempered (toughened) glass
- Chemically strengthened glass
- Laminated glass.

Float glass is produced by pouring molten glass onto a bed of molten tin by the process invented by Sir Pilkington in 1953. The glass floats on the tin and levels out as it spreads along the bath, giving a smooth face to both sides although the two sides of a glass sheet tend to be slightly different. Due to its molecular structure, the behaviour of glass is perfectly elastic until failure and there are no plastic deformations before the appearance of the first cracks in the structure. Float glass is produced in standard metric nominal thicknesses of 2, 3, 4, 5, 6, 8, 10, 12, 15, 19 and 25 mm in jumbo sheet stock sizes of 3.21 × 6 m [4]. Oversized jumbo sheets for the glass market are produced to a limited extent. For special purposes, some glass factories produce sheets up to 12 m in length [3]. Float glass is the most widely used type of glass today.

Annealed glass is basically float glass produced by a cooling process slow enough to avoid internal stresses caused by heat treatment in the glass. Glass can be made more load resistant by inducing the compressive stresses on the surface. It becomes annealed if heated above the transition point and then allowed to cool slowly. If glass is not annealed, it will retain many of the thermal stresses caused by quenching and will sustain a significant decrease in its overall strength [4]. Annealed glass is very brittle and breaks into large pieces (Fig. 4.2) which can cause serious injuries sustained by people being close to such glazing surfaces. Therefore, certain codes do not allow the use of annealed glass in areas where a

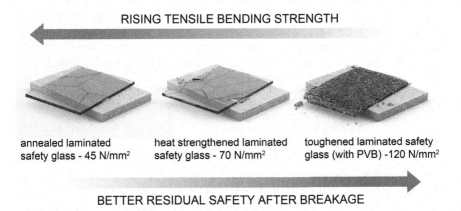

RISING TENSILE BENDING STRENGTH

annealed laminated
safety glass - 45 N/mm²

heat strengthened laminated
safety glass - 70 N/mm²

toughened laminated safety
glass (with PVB) -120 N/mm²

BETTER RESIDUAL SAFETY AFTER BREAKAGE

Fig. 4.2 Standard laminated glass types with their corresponding breakage forms and the tensile bending strength, adopted from Močibob [4]

risk of such injuries exists. Annealed glass has the lowest mechanical strength of all modern basic structural types of glass presented in this subsection.

Heat-strengthened (partially tempered) glass is the most common type of strengthened glass used in structural resisting elements. It has a thickness of less than 12 mm and has been tempered to induce surface residual stresses, but at a lower temperature and with a lower cooling rate than fully tempered glass. Hence, the name partially tempered glass. It differs from fully tempered glass in having lower residual stresses and breaking into evidently larger pieces, but still smaller ones than in the case of annealed glass (Fig. 4.2). Its tensile bending strength is halfway between the annealed (45 N/mm^2) and the fully tempered glass strength (120 N/mm^2), ranging at about 70 N/mm^2.

Fully tempered (toughened) glass is made from annealed glass by a thermal tempering process patented by R.A. Seiden. The process of manufacturing starts with glass being placed onto a roller table which takes it through a furnace that heats the glass to above its transition temperature. The glass is then rapidly cooled with draughts of air in a manner that lets the inner portion of the glass remain free to flow for a short time [4]. Fully tempered glass must be cut to the size and pressed to the shape before tempering, since glass once tempered cannot be reworked. This type of glass is referred to as safety glass because it breaks into small cuboid pieces, as opposed to ordinary annealed glass (Fig. 4.2), and reduces the risk of injuries caused by the breaking of glass panes. Toughened glass has consequently gained popularity in structural design. Its advantages taken from the structural point of view lie in the tensile bending strength reaching up to 120 N/mm^2, which is almost triple the value measured in ordinary annealed glass.

Chemically strengthened glass is a result of a process of strengthening by submerging glass into a bath containing a potassium salt or potassium nitrate heated to 450 °C. This causes sodium ions in the glass surface to be replaced by the larger potassium ions from the bath. Consequently, the potassium ions block the gaps left by the smaller sodium ions when these migrate to the potassium nitrate [4]. In contrast to fully tempered glass, chemically strengthened glass can be cut after the manufacturing process, but it losses the obtained additional strength within the area of about 20 mm from the cutting zone. Chemically strengthened glass cannot be classified as a safety glass because it breaks into long pointed pieces, similarly as annealed glass, and must be therefore laminated when applied in buildings. On the other hand, owing to chemical strengthening, the increased tensile bending strength of this type of glass is the highest and can reach the value of even 150 N/mm^2.

Laminated glass is not considered as a glass type but can be treated as a glass product composed of glass sheets glued together in a manner to improve the residual load-bearing capacity of glass panes. It was patented in 1903 by a French chemist, Eduard Benedictus. Laminated glass can be composed of annealed, partially tempered, fully tempered or even chemically strengthened glass sheets (Fig. 4.2). The majority of mechanical properties of laminated glass depend on the glass type of the sheets glued together with a transparent interlayer whose thickness is usually a multiple of 0.38 mm. The most commonly used interlayer is polyvinyl butyral (PVB), followed by cast-in-place resin (CIP), ethylene vinyl acetate (EVA) and SentryGlas Plus (SGP). Upon breaking of glass sheets, the interlayer holds the glass pieces together and assures a certain level of post-breakage resistance of the panes in addition to protecting the glass element against total collapse. As a result, laminated glass remains glued to the foil when shattered and has an increased residual load-bearing capacity. It is therefore used to ensure the resistance after breakage in areas submitted to a possible human impact, where glass could fall if shattered [4]. Specialist glass-processing companies are able to laminate single and multi-layer laminated sheets up to a jumbo panel size of 3.21 × 6 m, in exceptional cases even up to 7 m in length [3].

Breakage forms of the described standard glass types can be observed in Fig. 4.2. All structural glass types are presented as laminated glass products whose pieces are kept together on PVB after breakage. The level of safety, meant as a breakage form according to the size of the pieces, increases with the degree of strengthening, which is also true of the tensile bending strength whose values mount from 45 to 120 N/mm^2. Other mechanical properties, such as the compressive strength, the modulus of elasticity and the coefficient of thermal expansion, remain constant and do not depend on the strengthening of glass. The values for float glass given in Table 4.3 can be consequently adapted to all the presented types of structural glass.

4.2.2 Adhesives

The function of adhesives in timber-glass composites is to bind the two resisting materials—timber and glass whose mechanical properties show significant differences (c.f. Table 4.3 and Fig. 4.1). Glass is a very brittle material with practically no post-breakage capacity as opposed to timber which is more a ductile, but a very flexible material having a very low modulus of elasticity. It is consequently of utmost importance for adhesives to assure resistance and a high range of ductility of such composed load-bearing elements, simultaneously to finding balance between strength and deformability. Adhesives must also allow for expansion and shrinkage of timber, according to loading and humidity variations [6].

According to [7], adhesives used in timber-glass composites can be classified into three groups:

- Highly resistant and insufficiently flexible adhesives—rigid adhesives (acrylate, epoxy)
- Highly flexible adhesives, yet insufficiently resistant to loading—elastic adhesives (silicone)
- Adhesives that balance both key factors—strength and flexibility—semi-rigid adhesives (polyurethane, superflex polymers).

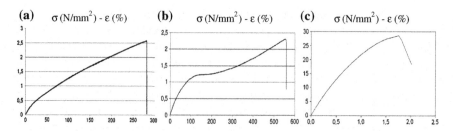

Fig. 4.3 σ-ε diagrams of different types of adhesives in tension; **a** silicone, **b** polyurethane, **c** epoxy

Stress–strain (σ-ε) diagrams in tension of all three basic adhesives types are schematically presented in Fig. 4.3.

The above diagrams obtained through short-term tests show an essentially higher strength of epoxy in comparison with polyurethane or silicone. What is more, the strains in epoxy can be even 100 times lower than those in silicone or polyurethane, which needs to be considered when deciding on the type of adhesive. Since we usually deal with the mid-term and long-term loads in practice, we ought to draw attention to findings by Haldimann et al. [8] who proved that the long-term strength of silicone is only about 10 % of its short-term strength due to a highly creep behaviour of silicone sealants. Other findings deserving to be mentioned are those by Cruz et al. [7], obtained from the shear tests results showing that the failure mode of timber-glass elements depends on the strength of the adhesive. It can be generally observed that glass regularly collapses in combination with high-resistance adhesives.

A further matter of importance is strong dependence of the optimal thickness of the bond line between timber and glass on the strength of the adhesive. The thickness of elastic adhesives is approximately 3 to 4 mm, while that of rigid adhesives ranges only from 0.3 to 0.5 mm. Figure 4.4 demonstrates strength behaviour of both adhesive types in dependence on the thickness of the bond line.

The main advantages and disadvantages of the above-mentioned adhesives used in timber-glass applications are thoroughly discussed in different studies, e.g., in Blyberg et al. [9] and Blyberg [10]. As seen in Fig. 4.4a, b, the results of the adhesive testing made within the above-listed studies showed that acrylate and polyurethane adhesives had significantly higher strength than silicone adhesives. Acrylate has a glass-transition temperature of 52 °C, which means that the properties of acrylate adhesives may undergo a substantial change when the temperature is increased. Winter et al. [11] claim that acrylates exhibit a dramatic reduction in strength when exposed to temperatures above 50 °C or to extreme humidity (RH). On the other hand, a study by Blyberg [10] on the effect humidity has on the acrylate adhesive bond did not indicate any huge effects on the strength of specimens kept at 85 % RH, which is a humidity level expected within indoor climate conditions.

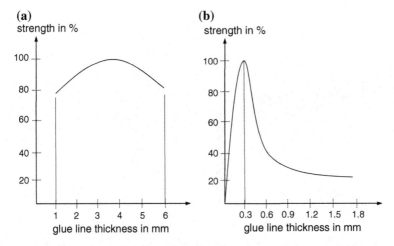

Fig. 4.4 Strength of elastic (**a**), and rigid (**b**) adhesives in dependence on the glue line thickness

Pequeno and Cruz [12] conducted a meticulous analysis of three different adhesives types (silicone, polyurethane and polymer), with respect to a number of structural and aesthetic aspects. The silicone adhesive proved to be the most advisable, as it allows for greater indexes of flexibility and assures the needed structural mechanical resistance. Moreover, silicone showed the highest UV resistance which is an utterly important fact to be taken into account when installing glazing surfaces in south-oriented façades, in view of highest possible degree of solar gains.

4.2.3 Insulating Glass

In order to understand the main functions of insulating glass and compare different glazing structures, basic knowledge of building physics is required.

4.2.3.1 Transmission of Solar Radiation Through Glazing

As solar radiation hits the glass panes, it is partly reflected, partly absorbed and partly transmitted directly through the glass. The absorbed radiation heats up the glass panes and is later emitted to the interior and exterior through heat radiation and convection (Fig. 4.5).

Figure 4.5 clearly shows that the total transmitted solar energy consists of directly transmitted solar radiation and of radiation absorbed in the glass panes and transformed into heat which is later emitted to the interior. The amount of total transmitted solar energy is expressed by the value of g, the coefficient of permeability of the total solar radiation.

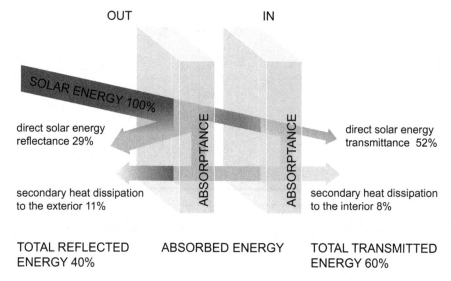

OUT   IN

SOLAR ENERGY 100%

direct solar energy
reflectance 29%

ABSORPTANCE

ABSORPTANCE

direct solar energy
transmittance 52%

secondary heat dissipation
to the exterior 11%

secondary heat dissipation
to the interior 8%

TOTAL REFLECTED   ABSORBED ENERGY   TOTAL TRANSMITTED
ENERGY 40%   ENERGY 60%

Fig. 4.5 Scheme of solar energy flow through an insulating glass unit

The concept of energy-efficient building design attributes a vital role to solar radiation, i.e., to solar gains through the transparent building envelope. Due to their solar radiation permeability, windows can contribute to the energy balance of buildings. On the other hand, widows represent areas in the building envelope with the highest heat loss potential, since the average U-value of windows is generally higher than the average U-value of opaque building elements (walls, ground slab and roof). However, constant development of the insulating glass technology results in launching ever new products with multiple gas-filled chambers and different coatings which substantially contribute to the reduction in heat losses through glass panes.

How is heat transferred through the insulating glass unit and what is the amount of heat flow influenced by? Heat transfer through a window occurs via three main mechanisms; conduction, convection and radiation (Fig. 4.6).

Conduction heat flow is transferred through adjacent atoms and molecules of gasses or solids. Heat always transfers from the warmer to the cooler side of a window, which means that the direction of conductive heat flow may change in the course of a day, month or year. Conduction heat flow occurs through the glass panes, the edge seal or spacer bar, the frame and even through the air or inert gas in the pane interspace. It can be minimized by adding glass panes, by using low-conductivity gasses, spacers with thermal brake and frames made of low-conductivity materials. Convection heat flow is the transport of heat away from the surface caused by air movement. It occurs in the pane interspaces and on each external side of the window. The use of inert gasses in the pane interspaces reduces energy losses due to convection. Radiation is a thermal exchange between the surface and the surrounding and always moves from a warmer surface to the cooler

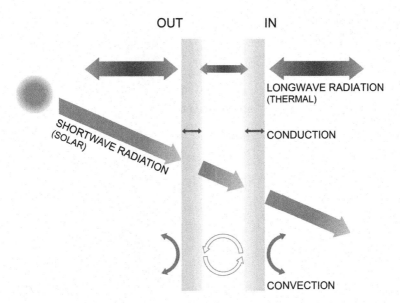

Fig. 4.6 Mechanism of heat transfer through the insulating glass unit

surrounding. Heat is usually radiated from the surface of the heated elements into the air and absorbed by glass to be reradiated afterwards to the interior or exterior. Radiation heat losses can be reduced by the use of low-emissive coatings applied to the glass panes.

The official definition of "insulating glass" is determined in EN 1279-1 [60] as "Multiple-pane insulation glass is a mechanically stable and durable unit comprising minimum two glass panes that are separated from each other by one or more spacing elements and are hermetically sealed at the edges". A standard insulating glass unit consists of two or three panes, while a high-efficiency insulating glass unit consists of four or even more glass panes.

4.2.3.2 Energy Indicators of Insulating Glass Units

A contemporary insulating window element consists of a glazing unit and window frame. The energy performance of windows can be expressed by two general indicators:

- The coefficient of thermal transmittance U [W/m^{2}K] with separate values for the glazing, the frame and the entire window element. It indicates the amount of heat of passing through 1 m^2 of component per unit of time based on a temperature difference of 1 K
- The coefficient of permeability of the total solar energy g [% or values 0–1] of the glazing. It indicates the sum of solar energy transmitted directly through the

Fig. 4.7 Main indicators of the energy efficiency of windows

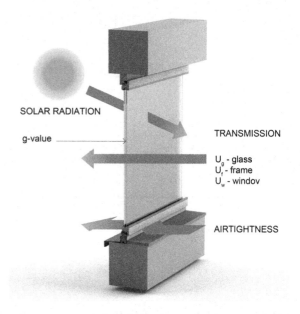

SOLAR RADIATION

g-value

TRANSMISSION

U_g - glass
U_f - frame
U_w - windov

AIRTIGHTNESS

glass and solar energy absorbed in the glass panes and later emitted to the interior.

Besides the above-listed indicators, proper airtight installation plays a vital part in the energy-efficient performance of windows, since it significantly reduces infiltration causing ventilation heat losses (Fig. 4.7).

Glazing properties exert influence not only on the building's thermal performance but also on the quality of the interior daylight. Sufficient amount of visible radiation transmitted through the glazing reduces the need for artificial lighting, which saves electrical energy. Indicators expressing the quantity of transmitted and reflected light are the following:

- The light transmission coefficient LT [%], expressing the percentage of visible solar radiation (wavelength from 380 to 780 nm) transmitted through glass.
- The light reflection coefficient R [%], expressing the percentage of visible solar radiation (wavelength from 380 to 780 nm) reflected by glass.
- Additional indicators showing the quality or quantity of transmitted solar energy are as follows:
- Colour rendition index Ra [0–100], indicating colour recognition in the interior and colour recognition through the glazing itself. The highest value of 99 indicates neutral colour recognition.
- The selectivity factor $S = LT/g$, representing the ratio between light transmittance LT and the degree of total solar energy permeability g. A higher S-value expresses a better ratio.

- The shading coefficient Sc = $g/0.80$ indicates the mean transition of solar energy related to the degree of total solar energy transmittance of an uncoated two-pane insulated glass unit. This indicator is essential for calculating the required cooling load of the building [UNIGLAS® [13]].

In addition to the main indicators expressed above, glass producers use a number of other indicators to provide accurate description of their products, i.e., of the quantity or quality of the solar energy absorbed, transmitted or reflected by the glass panes.

4.2.3.3 Parameters Influencing Energy Properties of Insulating Glass

A window is treated as a complex element with its overall coefficient of thermal transmittance (U_w) depending not only on the glazing but also on the type of window frame and spacer element. The U_w can be read or measured or calculated. Determination by means of calculation is based on the following equation according to ISO EN 10077-1:2006 [14]:

$$U_w = \frac{A_g \cdot U_g + A_f \cdot U_f + l_g \cdot \psi}{A_g + A_f}$$ (4.1)

with the following quantities presented in Fig. 4.8:
U_w The heat transfer coefficient of the window,
U_f The heat transfer coefficient of the frame,
U_g The heat transfer coefficient of the glazing,
A_f The area of the frame,
A_g The area of glass,
l_g The length of the glazing perimeter,
ψ Linear heat transmittance of the glass edge (describes thermal bridges of a constructional component).

Fig. 4.8 Areas of the insulating window considered in calculation of the U_w

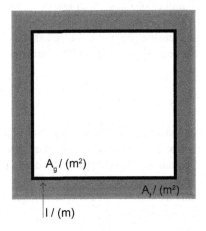

$A_g / (m^2)$

$A_f / (m^2)$

$l / (m)$

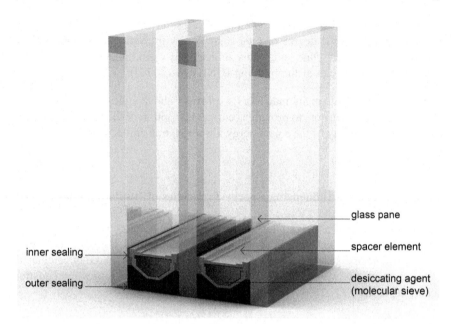

glass pane

spacer element

inner sealing

outer sealing

desiccating agent
(molecular sieve)

Fig. 4.9 Composition of a typical three-pane insulating glass unit

Thermal transmittance (U_g) of the glazing depends on the number and dimension of the pane interspaces, on the type of gas filling, on the number, type and emissivity of the heat insulation layers, i.e., glass coatings, and on the type of edge seal and spacer element. The composition of a three-pane insulating glass unit is presented in Fig. 4.9.

Number of Glass Interspaces

Increasing the number of glass panes and pane interspaces results in the reduction in heat flow and light transmittance.

Glass Coatings

A major part of the energy loss in an insulating glass unit is caused by radiation, while in a window frame, the main energy loss is due to conduction. As a consequence, different parts of the window require different approaches to energy loss reduction. In the area of glazing, the main energy loss reduction strategy should therefore focus on preventing heat to radiate from the interior to the exterior.

The above-mentioned heat radiation prevention can be achieved by using glass coatings, i.e., thin metal and metal oxide layers which are applied to the glass surface either by the pyrolytic or the magnetronic method with a purpose of reducing the permeability of radiation through the glass panes. It depends on the selectivity of the coating whether the latter reduces short-wave solar radiation (solar coatings) or long-wave heat radiation (heat coatings or low-e coatings). In the pyrolytic method, molten metal or metal oxide is applied to glass at high temperatures by either dipping or spraying, which results in a hard coating. Magnetronic sputtering is a process of applying thin layers of various metal and metal oxide coatings to glass as soft coatings. A wide range of light reflection, light transmission, infrared reflection and colour options of the glass surface are thus permitted. Magnetronic sputtering also allows for a combination of heat and solar coatings to be applied—one on top of the other [15].

Individual layers of heat coatings (also called low emissivity or low-e coatings) are invisible and act as selective filters which are permeable to the short-wave solar radiation of a wave length up to 2,500 nm, but impermeable to the long-wave heat radiation, especially the long-wave IR radiation of the wave lengths ranging from 3,000 to 50,000 nm. As seen from a practical viewpoint, solar radiation can pass through the glass into the room where it is absorbed by the surfaces of the interior which warm up and later emit the energy as the long-wave IR radiation. Heat coatings applied to glass act as impermeable to the IR long-wave heat radiation by reflecting the radiation back to the room. Low-e coatings are usually applied to the glass pane adjacent to the interior to the side facing the cavity or potentially to both outer panes to the sides facing the cavity. The panes' surfaces are normally numbered from the outside to the inside (Fig. 4.10), which helps describing the position of individual coatings. Application of low-e coatings reduces the U_g-value.

Solar coatings reduce the permeability of short-wave solar radiation through the glass panes, which results in the lower g-value for glass. The extent to which the

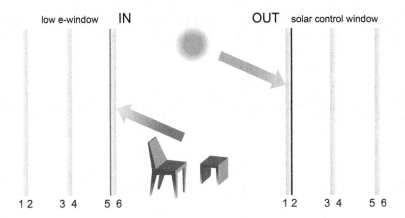

Fig. 4.10 Basic principle of positioning the coatings in the insulating glass unit

g-value and the light transmittance of glass are reduced depends on the ratio of selectivity of the solar coating. Almost all solar coatings, with the exception of soft coatings reduce both, the permeability of solar energy and the light transmittance. Solar coatings are usually applied to the outer pane adjacent to the exterior to the side facing the cavity. Some hard coatings can be applied to the external side of the outer glass pane (position 2 in Fig. 4.10), since they are resistant to environmental conditions. It is also possible to apply a combination of both, solar and low-e coatings on a single-glazing pane.

Where and why do we use solar coatings? In residential buildings, the strategy of passive solar design usually refers to exploitation of solar energy transmitted through transparent surfaces in order to reduce the energy demand for heating. On the other hand, there exists a risk of overheating in the summer period, which can be prevented by the use of appropriate shading elements. The use of solar coatings for the glazing in residential buildings is not advisable, since it has a negative impact on the principle of solar energy use and the quality of daylight. There are nevertheless some exceptions in residential architecture, e.g., large winter gardens, pavilions, glazed swimming pools, where the prevention of summer overheating is more important than the solar energy gain in winter. In the architecture of public, office and administration buildings, a permanent trend heading towards transparency requires large glazing surfaces. Since the use of external shading elements is not always appropriate in such public buildings, the use of sun protective coatings reduces overheating of the rooms by means of reflection and absorption of solar energy and a consequent reduction in the load on air conditioning systems in buildings.

Number of Glass Interspaces and Cavity Fillings

Increasing the number of glass panes and pane interspaces causes the reduction in heat flow and light transmittance. The glazing interspace in a standard insulating glass unit is usually 12–16 mm wide and filled with a rare gas in order to reduce thermal conductivity. The gas used in most cases is argon or krypton. The influence of the type of cavity gas of the number of pane interspaces and low-emissive coatings on the U_g-value is presented in Table 4.4. The data are based on the review of Gustavsen et al. [16].

Table 4.4 Typical U_g-values for different configurations of the window glazing

Glazing U_g-value [W/m²K]			
Glazing configuration [mm]	Cavity gas		
	Air	Argon	Krypton
No coating	4	5.8	–	–
	4-12-4	2.9	2.7	2.6
	4-12-4-12-4	2.0	1.9	1.7
Low-emissive coating	4-12-E4	1.6	1.3	1.1
	4-12-4-12-E4	1.3	1.0	0.8
	4E-12-4-12-E4	1.0	0.7	0.5

According to the above data, application of multiple interspaces, gas fillings and coatings into the configuration of a glass unit reduces the U_g. Data based on the review of Gustavsen et al. [16] show that a 4 mm pane of float glass has a U_g of approximately 5.8 W/m^{2}K. A double-pane insulating unit without coatings filled with argon has a U-value of approximately 2.7 W/m^{2}K. Adding a single low-emissive coating leads to a reduction in U_g which then attains a value of 1.3 W/m^{2}K. In a triple-glazing unit, filled with argon and coated on both external panes, the U_g is about 0.7 W/m^{2}K, while if filled with Krypton, the U_g-value approximates 0.5 W/m^{2}K. Insulating glazing units currently under development are made of multiple chambers with different coatings and gas-filled cavities whose values range around $U_w = 0.30$ W/m^{2}K, which is relatively close to the U-value of external walls prescribed for low-energy houses ($U = 0.20$ W/m^{2}K).

Spacer

Spacer is an element that holds the panes in an insulating glass unit at the appropriate distance. To prevent condensation in the pane interspaces, the spacer is filled with a desiccating agent, which can absorb any penetrating moisture. In addition, it has to assure the sealing of the air- or gas-filled pane interspace which is made of two seals, a primary seal between the spacer profile and the glass and the secondary seal applied to the external side of the spacer bar. Many different types of spacers can be currently found on the market. The weakness of the standard aluminium spacers is their high conductivity resulting in a large thermal bridge causing the cold edge effect in the edge area of the insulating glass unit, which increases a risk of condensation. The market nowadays offers several innovative edge seal types with lower conductivity and interrupted thermal bridging in the edge area (warm edge effect), which improves the U-value of the entire insulating glass unit and significantly contributes to the increase in the interior surface temperature at window edges. Such edge seals include thermally broken aluminium or stainless steel spacers, silicone foam spacers, corrugated metal spacers, fiberglass spacers, PVC spacers, etc. Gustavsen et al. [16]. With the increasing use of high-insulating windows, a thermally improved edge spacer has become an important element.

Window Frame

The thermal performance of the window frame has an effect on the thermal performance of the entire window. Depending on the façade type, there exist different manners of installation of the insulating glass units, such as insulating glass units with uncovered edge seals or those with mechanically retained edge seals or insulating glass units mounted in a frame which completely covers the edge seal on all four sides. The selection of the window frame often depends on the architectural design concept. However, the material of the window frame, which normally has a higher U-value than the glazing unit, must be carefully selected. The choice can be made among wood frames, PVC frames, fiberglass frames, aluminium and steel frames with a thermal brake or composite frames made from an internal wood frame covered with external PVC or insulation-filled aluminium cladding. The U_f-value of wood frames is mainly decided by the thickness of the casement and the frame in the heat flow direction. There exist also wood frames with a number of air cavities, which additionally reduce the U_f-value. For PVC frames, the U_f-value mainly depends on the number of air cavities and the location of the load-carrying element which is usually made of metal. Window frames made of metal, usually aluminium, should be constructed of an outer and an inner profile and separated by an insulating material (a thermal break), i.e., polyamide [16]. Wood frames generally have a lower U_f-value in comparison with aluminium or plastic frames. Moreover, lower values of embodied energy make wood frames more sustainable than aluminium, PVC, steel and aluminium-cladded wood types of frames.

Other methods used for the regulation of thermal and optical performance of windows.

Apart from the parameters described above, a set of other methods can be used in order to adjust optical and thermal behaviour of windows. Contrary to the approaches already described, these methods are not static since they use either electrical voltage or chemical substances to achieve a temporary change in the light and energy transmittance. Currently, different products such as electrochromic or gasochromic windows are already either being launched or expected to reach the market in the near future. Electrochromic windows darken when voltage is added and become transparent when voltage is taken away. Such windows can be adjusted to allow for varying levels of visibility. Gasochromic windows produce a similar effect as electrochromic windows, but in order to colour the window, diluted hydrogen is introduced into the pane interspace. Upon adding oxygen, the coated surface bleaches and the window returns to its original transparent state. These contemporary products also known as *switchable glass* or *dynamic glass* represent a great potential for the use in large-size glass façades.

4.3 Research Related to the Optimal Glazing Size and Building Shape

Researching energy efficiency of buildings is not solely a matter of the last decade, as the first intensive studies related to energy and buildings date from the seventies and eighties of the previous century. Many studies focusing on the research of specific parameters that exert influence on the energy performance of buildings, such as Johnson et al. [17] and Brown [18], have been performed since then. Johnson et al. [17] systematically explored the influence of the glazing systems on the component loads and the annual energy use in office buildings typical of different climates and orientations. Another research estimating the total area of the exposed surface of domestic buildings carried out by Steadman and Brown [18] involved an empirical study of a house plan drawn for a building in the city of Cambridge. Within a range of researched parameters, such as the relationship between the wall and the floor area, the built form, the glazing areas were examined from the viewpoint of heat loads. One of the comparable newer researches is also a parametric case study of an apartment building, fictively located in five different Turkish cities, presented by Inanici and Demirbilek [19]. The effects of variable parameters on the annual energy consumption, such as different building's aspect ratios and different south window sizes, were analysed in order to determine the optimal parameter values. Next, a parametric study of the heating and cooling demand was performed by Bülow-Hübe [20] in order to determine an optimal design for office windows in Swedish climate. Another Swedish study for 20 low-energy terraced houses built in 2001 outside Gothenburg was performed by Persson et al. [21]. The purpose of the work was to investigate how decreasing the south-oriented window size and increasing the north-oriented window size could influence the energy consumption. A number of findings are furthermore stated in the dissertation by Persson [22]; these are in certain aspects comparable to our research. In the framework of a European project Ford et al. [23], various simulations and analyses were performed for different low-energy buildings for five European countries (UK, France, Italy, Portugal, Spain) with relatively warm climates. Many of the existing studies were carried out for non-European climates. Bouden [24] investigated the appropriateness of glass curtain walls for the Tunisian local climate. The influence of windows on the energy balance of apartment buildings in Amman, Jordan, was analysed in the study performed by Hassouneh et al. [25]. In a study, focusing on the role of active systems and thermal protection in passive and plus energy residential buildings [26] explain that south-oriented windows make the solar gains outweigh the heat losses by 2/3. The relationship between the solar gains and heat losses is almost equal for south- and west-oriented windows; while for north-oriented windows, the heat losses are about 3 times higher than the solar gains.

In general, all of the presented studies deal primarily with the influence of variable parameters on the energy performance of different types of buildings (residential, offices, public) of mainly massive construction systems. From the existing research findings, we summarize that the process of defining the optimal model of a building is very complex. The most important parameters influencing the energy performance of buildings are listed below:

- Location of the building and climate data for the specific location
- Orientation of the building
- Properties of the materials installed, such as timber, glass, insulation, boards
- Building design (shape factor, length-to-width ratio, glazing size, building envelope properties, window properties)
- Selection of active technical systems.

It is therefore important to investigate the influence of the above-listed parameters with utmost care. Due to the absence of a direct correlation between the different parameters, it is more convenient to conduct separate examinations of their influence on the energy demand for buildings. The latter is of particular relevance to the influence of the building's orientation and its glazing size which will be a subject of common analysis in Sect. 4.3.1 on the one hand, and to the influence of the building shape which will be thoroughly examined in Sect. 4.3.2, on the other.

However, it is important to stress that the presented calculations do not consider various active systems' impacts (heat recovery ventilation, solar collectors, PV panels, heat pumps, etc.). The results of the comparative analysis can nevertheless serve as a good frame of reference to architects and civil engineers in an approximate estimation of the energy demands accompanying different positioning and proportion of the glazing surfaces, while using various prefabricated timber-frame wall elements.

4.3.1 Influence of the Glazing Arrangement and its Size on the Energy Balance of Buildings

One of our general critical remarks referring to the existing studies focusing on the impact of windows on the heating and cooling demand was that most of them are just calculations for a single building. In our research, an attempt at a more systematic analysis was made, with the model of a building of our base case study being performed in many variations of timber construction systems. The first part of the study presents a parametric analysis of the glazing-to-wall area ratio impact in a two-storey house with a prefabricated timber-frame structural system. The analysis was carried out for different construction systems and for different cardinal directions. Based on a parametric analysis, the second part presents a generalization of the problem related to the energy demand dependence and to the optimal glazing area size dependence on one single variable, the U_{wall}-value, which becomes the only variable parameter for all contemporary prefabricated timber construction systems, independently of their type. Finally, mathematical linear interpolation is presented as a simple method for predicting an approximate energy demand with regard to the glazing size and the U_{wall}-value in the analysed case study, thoroughly presented in Žegarac Leskovar and Premrov [27].

Among the parameters listed previously in this chapter, our case study examines the influence of the following three: the glazing-to-wall area ratio, the U_{wall}-value and the main cardinal directions for a specific climate. Since the current study limits itself solely to timber construction, which is also termed as lightweight construction, the influence of different thermal capacities of the building materials was not taken into consideration. The presented approach could be also applicable to massive construction (brick, concrete walls), if additional parameters concerning thermal mass were considered, although we would expect slightly different results in the case of a building model in a massive construction system. Calculations do not consider various active systems impacts (heat recovery ventilation, solar collectors, PV panels, heat pumps, etc.), or various window U-value impacts.

4.3.1.1 Parametrical Numerical Study

The current subsection presents a parametric numerical case study of a two-storey house and its parametric analysis of the glazing-to-wall area ratio impact on the energy demand. A model was selected out of sixteen projects created in the study workshop "Timber Low-Energy House". The workshop was held upon a public call made by the Slovene timber house manufacturers. The aim of the workshop was to develop different innovative models of timber-glass low-energy houses, suitable for a typical European family of 4–5 members. Consequently, the number of occupants planned for the purposes of our parametric study was 4. The external horizontal dimensions of the model are 11.66×8.54 m for the ground floor and 11.66×9.79 m for the upper floor (Fig. 4.11). The total heated floor area is 168.40 m^2 and the total heated volume is 437.80 m^3.

The three-dimensional model of the house is presented in Fig. 4.12.

Construction

The exterior walls are constructed using a timber-frame macropanel system. All the analysed wall elements are vertical. The exterior wall U-value is 0.102 W/m^{2}K for the TF-3 element (c.f. Table 3.6 and Fig. 3.36c). The U-values of other external

Fig. 4.11 Floor plans of the base—case study model

Fig. 4.12 Three-dimensional
model of the house

construction elements are 0.135 W/m²K for the floor slab, 0.135 W/m²K for the flat
roof and 0.130 W/m²K for the south-oriented overhang construction above the
ground floor area. The composition of the basic TF-1 and the thermally improved
TF-2 macropanel wall element is listed in Table 3.6 and presented in Fig. 3.36a, b.

Additional modifications of AGAW are made only for the south-oriented
glazing areas for two classical single-panel systems TFCL-1 and TFCL-2 (Fig.
3.34a) with higher U-values. The composition of the treated single-panel timber-
frame construction systems is presented in Table 4.5. The composition of the
fictive single-panel wall system TFCL-3 with the U_{wall}-value of 0.30 W/m²K is
taken from Žegarac Leskovar [28] and Žegarac Leskovar and Premrov [27].

Glazing

A window glazing (Unitop 0.51–52 UNIGLAS) with three layers of glass, two
low-emissive coatings and krypton in the cavities with a configuration of 4E-12-4-
12-E4, is installed. The glazing configuration with a g-value of 52 % and
$U_g = 0.51$ W/m²K assures a high level of heat insulation and light transmission.
The window frame U-value is $U_f = 0.73$ W/m²K, with the frame width being
0.114 m. The glazing-to-wall area ratio (AGAW) of the south-oriented façade is
27.6 %, with the AGAW values of the rest of the cardinal directions being 8.9 %
in the north-oriented, 10.5 % in the east-oriented and 8.5 % in the west-oriented
façades.

Climate and Orientation

The house with a large glazing area installed in its longer side facing south is
located in Ljubljana. The city of Ljubljana is located at an altitude of 298 m, a
latitude of 46°03′ north and a longitude of 14°31′ east. According to the accessible
climatic data from ARSO [29], the considered average annual external temperature
is 9.8 °C. The relevant climate data are listed in Table 4.6.

Table 4.5 Composition of the analysed single-panel wall elements

TFCL-1		d [mm]	TFCL-2		d [mm]
Material			Material		
Wooden planks		22	Wooden planks		22
TSS* with open air gaps	Bitumen sheet cardboard	0.5	TSS with open air gaps	Bitumen sheet cardboard	0.5
TSS with open air gaps	Timber frame	50	TSS with open air gaps	Timber frame	20
Bitumen sheet cardboard	Timber frame	0.5	Bitumen sheet cardboard	Timber frame	0.5
MW	Timber frame	50	MW	Timber frame	80
Aluminium foil			Aluminium foil		
Particleboard		13	Particleboard		13
Gypsum plasterboard		10	Gypsum plasterboard		10
total thickness [mm]		146	Total thickness [mm]		146
U_{wall}-value [W/m^{2}K]		0.70	U_{wall}-value [W/m^{2}K]		0.48

*Timber substructure, **Mineral wool

Table 4.6 Annual climate data for Ljubljana [29]

	Annually
Average temperature	9.8
Average relative humidity at 7 am (%)	90.2
Average relative humidity at 14 pm (%)	62.4
Average duration of solar radiation (h)	1,712
Nr. of clear days (cloudiness < 2/10)	32.5
Nr. of cloudy days (cloudiness > 8/10)	142.2
Nr. of days with fog	120.8

Shading

The house is constructed with a south-oriented extended overhang above the ground floor, which blocks direct solar radiation from entering the ground floor windows during the summer, while it lets enter in winter when the angle of incidence of the sun is lower. The rest of the windows on the upper floor and those of the east- and west-oriented walls are shaded with external shading devices.

Internal Gains and HVAC

The house is equipped with a central heat recovery unit. To prevent overheating in the summer period, night ventilation with cooling through manual window is planned. The interior temperatures are designed to reach a T_{min} of 20 °C and T_{max} of 25 °C. Domestic hot water generation (DHW) and an additional requirement for space heating are covered by a heat pump with a subsoil heat exchanger and, to a minimal extent (5 %), by electric heating.

Variable Parameters

The influence on the energy demand of the following factors is studied: the glazing size in four different cardinal directions: south, north, east and west. Modifications of the glazing area size are performed in the range of AGAW from 0 % to nearly 80 % (Fig. 4.13), separately for each cardinal direction, for three timber-frame

Fig. 4.13 South-oriented façade of the base—case model with the schemes of AGAW modification

macropanel systems: TF-1, TF-2 and TF-3. Modifications of the glazing area size are made step by step through adding window elements (frame + glazing) to the totally unglazed façade as presented in Fig. 4.13.

Description of the Software and the Calculation Method

The Passive House Planning Package [30] is used to perform static calculations of the energy demand. The software, certified as a planning tool for passive houses, allowing a surprisingly accurate description of thermal building characteristics of passive houses, can be also used for low-energy house design. Practice has shown that the results achieved by the PHPP software are very similar to the measured energy demand in operating buildings. It is important to stress that the main point of our study is to present an approach to optimal design. The selection of software is therefore not decisive, since many other calculation tools could be used as well. The calculation method of the parametric numerical case study process is graphically presented in Fig. 4.14. As shown, more than 160 calculations are made in order to obtain the results showing the effects of the selected parameters on the energy demand for heating and cooling. An upgrade of the calculation procedures

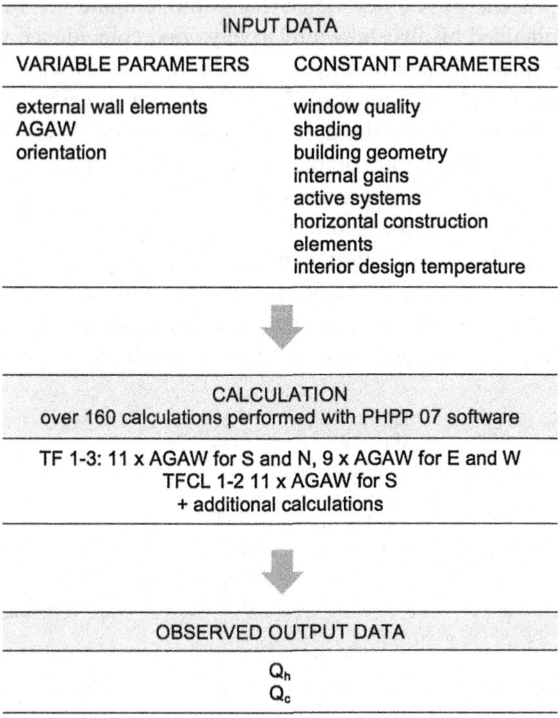

Fig. 4.14 Scheme presenting the calculation method

INPUT DATA	
VARIABLE PARAMETERS	**CONSTANT PARAMETERS**
external wall elements AGAW orientation	window quality shading building geometry internal gains active systems horizontal construction elements interior design temperature

CALCULATION
over 160 calculations performed with PHPP 07 software

TF 1-3: 11 x AGAW for S and N, 9 x AGAW for E and W
TFCL 1-2 11 x AGAW for S
+ additional calculations

OBSERVED OUTPUT DATA

Q_h
Q_c

will be presented later in this paper with the generalization of the results for one single-variable parameter (U_{wall}-value).

4.3.1.2 Results and Discussion

Figure 4.15 shows a comparison of the annual energy demand for heating (Q_h) as a function of the glazing area size for different cardinal directions of the TF-3 construction system with the lowest U_{wall}-value.

The results show evidence of the strongest influence of increasing the glazing area size in the south orientation where Q_h decreases almost linearly with a growing AGAW and the heat gains at AGAW = 0.79 add up to almost 52 % of the Q_h value at AGAW = 0. The increase in Q_h for almost 37 % related to the energy demand for heating at the starting point shows that the influence of altering the glazing area size facing north is less expressive than that of its southern counterpart. The east and west orientations, on the other hand, show almost identical behaviour.

The presented analyses generally accord well with the results of the parametric study research on the effect of the glazing type and size on the annual heating and cooling demand for the Swedish timber-frame offices [20] and low-energy houses [21, 22]. Taking into account the differences in climate, there is considerable agreement noticed with certain statements from design guidelines for comfortable low-energy homes considering the climate in Milan [23]. Furthermore, the obtained results show a relatively good coincidence with the values for the energy demand related to different glazing area sizes with different glazing types for the case study in Amman [25], with respect to certain differences in the external air temperature and the duration of solar radiation considered in the calculations.

The behaviour of the energy demand patterns of the TF-1, TF-2 and TF-3 systems for the west and east directions is very similar, while the patterns for the north orientation show only the increase in the energy demand. No noticeable decrease in the energy demand, either for Q_h or Q_c, appears for the north, west or east orientations. Therefore, only the south direction, the focal point of our special interest, is additionally analysed and compared for all construction systems. The most interesting point is the comparison of the sum total of the energy demand

Fig. 4.15 Annual energy demand for a heating (Q_h) in the TF-3 construction system as a function of AGAW for different cardinal directions

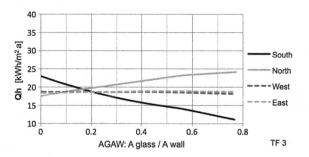

$(Q_h + Q_c)$ for different construction systems (TF-1 to TF-3), presented in Fig. 4.16. In the case of timber buildings, particular attention should be paid not only to the energy demand for heating, but to that for cooling as well. Due to a low thermal capacity of timber, the risk of overheating is considerably higher than in buildings made of brick or concrete.

The results for the sum total of the energy demand show an interesting phenomenon related to the optimal point with the lowest $Q_h + Q_c$ demand which is clearly evident in the TF-3 construction system, appearing at the range of AGAW $\approx$ 0.34–0.38, quite evident in the TF-2 system at AGAW $\approx$ 0.41 and less evident in the TF-1 system at AGAW $\approx$ 0.42–0.50. We assume that the optimal share of the glazing surface in the south-oriented exterior walls depends on the thermal transmittance of the exterior wall. The optimal share of the glazing area in walls with extremely low U-values is smaller than in walls with higher U-values. If we pay attention to the behaviour of the $Q_h + Q_c$ function curve after reaching the optimal point, we notice that the sum total of the energy demand for heating and cooling increases more in the TF-3 construction system, which has the lowest thermal transmittance, while in the TF-1 system with a higher U_{wall}-value, the function converges. The higher the U_{wall}-value of a specific system, the higher the values of the functional optimum.

It is interesting to compare the results with the study performed by Inanici and Demirbilek [19], who showed that in the process of increasing the south-facing window area, each increase in the glazing size led to a decrease of the total energy load $(Q_h + Q_c)$ for cool climates, while the opposite was true of hot climates. Due to a method of keeping a constant overall U_{wall}-value, a direct comparison with our case is not possible, although the analysis for Ankara, whose average annual temperature is similar to that of Ljubljana, showed interesting results when an additional calculation method was used. The lowest energy loads were shown at the maximal U_{wall}-value and the glazing-to-wall area ratio of 30 %, which is similar to our case.

For comparison purposes as well as for support in setting up the basic principle of the glazing surface impact on the energy behaviour patterns, an analysis of the classic single-panel prefabricated wall elements is carried out, but only for the south orientation. Firstly, TFCL-2 and an additional fictive wall element TFCL-1

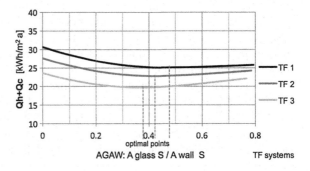

Fig. 4.16 Comparison of the sum total of the energy demand for heating and cooling as a function of AGAW for the south orientation of the selected TF construction systems (TF-1 to TF-3)

Fig. 4.17 Comparison of the energy demand for heating and cooling as a function of AGAW for the south orientation of the selected TF construction systems

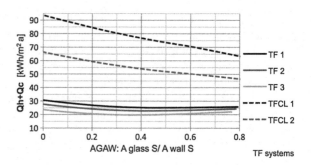

are analysed. Thermal properties of the selected wall elements do not satisfy even the basic requirements for the thermal transmittance of an exterior wall (U_{wall}-value < 0.20 W/m^{2}K for a lightweight construction) in low-energy house design. The analyses of the sum total of the heating and cooling demand presented in Fig. 4.17 seem to be most interesting.

It is evident from the presented results that at higher U_{wall}-values of the exterior wall elements, the functional optimum (the lowest $Q_h + Q_c$ value) disappears, the $Q_h + Q_c$ function curve passes from parabolic dependence in construction systems with extremely low U_{wall}-values (TF-2 and TF-3) to linear dependence in construction systems with high U_{wall}-values (TFCL-1 and TFCL-2). The inclination of the function line presenting TFCL systems depends on the U_{wall}-value. Energy decrease caused by an increase in the total glazing area (measured from AGAW $= 0$ to AGAW ≈ 0.80) represents approximately 33 % of the starting point value for the TFCL-1 system, but only 17 % for the TF-3 system with the highest insulation features (measured from AGAW $= 0$ to AGAW$_{opt}$).

4.3.1.3 Generalization of the Problem to One Single Independent Variable (U_{wall}-value)

Determination of AGAW$_{opt}$

The main aim of the presented study is in developing a theoretical approach applicable to architectural design of an optimal energy-efficient prefabricated timber-frame house. It is thus important to transform this complex energy-related problem, dependent on the structural system, to only one single independent variable (U_{wall}-value) which becomes the only variable parameter to determine the optimal glazing area size value (AGAW$_{opt}$) for all contemporary prefabricated timber construction systems.

The first step to be taken in setting up the basic theory of the research focusing on a single independent variable is to observe and compare the energy demand behaviour for both, the new macropanel wall elements and the classic wall elements with single-panel construction, where the thermal transmittance of the selected wall elements is fictively set at an equal value. In Fig. 4.18, we present the comparison of

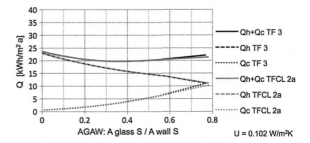

Fig. 4.18 Comparison of the energy demand as a function of AGAW for the south orientation of the selected TF-3 and TFCL-2a construction systems with a uniform U_{wall}-value = 0.102 W/m²K

the sum total of the energy demand ($Q_h + Q_c$) for TF-3 and TFCL-2a construction systems, where the wall elements with an equal U_{wall}-*value* = 0.102 W/m²K are analysed. The U_{wall}-value for TFCL-2a system is obtained by adding fictive insulation to the single-panel TFCL-2 wall element.

For the benefit of further approach, it is important to notice that the results presented in this particular case are almost equal for both construction systems. Additionally, we also analysed three different massive panel CLT systems (types KLH-1, KLH-2 and KLH-3) described in Sect. 3.2.2.2 and schematically presented in Fig. 3.22. The whole analysis with the calculated results is presented in Žegarac Leskovar [28]. The calculated results for the optimal AGAW values of all the analysed types of external wall elements are seen in Table 4.7.

Based on the results presented, it is now possible to analyse the relationship between the optimal glazing size in the south-oriented external wall elements (AGAW$_{opt}$) related to the $Q_h + Q_c$ energy demand and the thermal transmittance of the wall element (U_{wall}-value). The data presented in Fig. 4.19 show the values of AGAW, at which the $Q_h + Q_c$ demand reaches the lowest value, depending on the U-value of the external wall element as the only independent variable.

Figure 4.19 shows that the optimum or the convergence of the function curves for AGAW$_{opt}$ appears only in systems with a U_{wall}-*value* ≤ 0.193 W/m²K. As the U_{wall}-value increases, the optimal share of south-oriented glazing size becomes higher. Upon reaching the limiting U_{wall}-*value* = 0.193 W/m²K, the values of the optimal AGAW converge towards the maximal glazing surface. No optimum or convergence for AGAW appears in the analysed construction systems with an

Table 4.7 Optimal values of AGAW in south-oriented external wall element for selected timber construction systems

Construction system	U_{wall} [W/m²K]	AGAW$_{opt}$	AGAW$_{opt}$ adjusted
TF-1	0.164	0.42–0.50	0.47
TF-2	0.137	0.41	0.41
TF-3	0.102	0.34–0.38	0.37
KLH-1	0.181	0.52–0.54	0.53
KLH-2	0.148	0.41–0.46	0.43
KLH-3	0.124	0.38–0.40	0.39
Systems	≥0.193	≈0.80	0.80

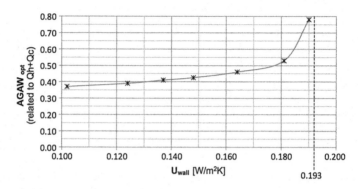

Fig. 4.19 Optimal values of AGAW in the south-oriented external wall element as a function of the U_{wall}-value for timber construction systems

U_{wall}-value > 0.193 W/m^{2}K. Although the lowest $Q_h + Q_c$ is reached at the maximal AGAW value, we have to pay attention to the data for the overheating frequency.

In conclusion, it is important to stress that the existing literature offers only predictive and in many cases even clearness-lacking parametric studies on the optimal glazing size depending on thermal characteristics of the external wall. Due to a large list of parameters affecting the energy performance of a building, it is important to assure a way to a fast and simple pre-estimation of the energy load. In contrast, the scientific contribution of the current study is presented by analytical functional dependence of AGAW on the U_{wall}-value (Fig. 4.19), which allows for a selection of any external wall element with a specific U_{wall}-value and consequently for a selection of an optimal AGAW value, which increases with the increase in the U_{wall}-value.

The presented generalization concerning the U_{wall}-value as the only variable parameter can be applied to timber construction in general, regardless of the construction system. The determined function for the optimal south-oriented glazing size (AGAW$_{opt}$) offers an opportunity to select an optimal renovation process that would combine the improvement of thermal properties of the external walls through an additional layer of insulation, with the installation of the optimal glazing size which is noticeably smaller in the case of a lower U_{wall}-value.

4.3.1.4 Use of the Linear Interpolation for a Simple Pre-estimation of the Energy Demand

In order to set the next general principle, a graphical presentation (Fig. 4.20) of the sum total of the energy demand for heating and cooling as a function of the U_{wall}-value for the three selected AGAW values (AGAW $= 0$, AGAW ≈ 0.40, AGAW ≈ 0.80) is of great importance. The results of the parametric numerical case study (Fig. 4.20) feature the $Q_h + Q_c$ demand for three glazing-to-wall ratios,

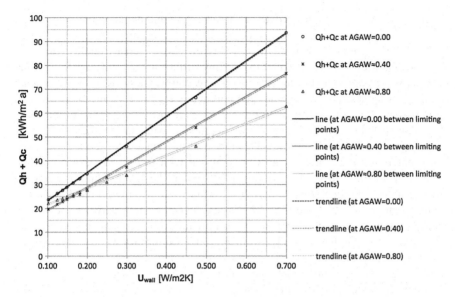

Fig. 4.20 $Q_h + Q_c$ demand as a function of the U_{wall}-value for three different AGAW values with possible ways of linearization

for AGAW $= 0$, AGAW ≈ 0.40 and for AGAW ≈ 0.80, along with possible ways of linearization.

The $Q_h + Q_c$ demand increases almost linearly with the increase in the U_{wall}-value. The values for the sum total of the energy demand for all three AGAW values are almost straight lines with an inclination angle depending on the AGAW value. Another item of importance is seen in the fact that the linear approximation between the two limiting selected points of the U_{wall} produces nearly the same results as the straight line with the smallest deviation. The process of calculation can thus become substantially shorter and simplified.

The linearization of the energy loads ($Q_h + Q_c$) problem shows that the use of linear interpolation is very simple and applicable, owing to the fact that it includes the influence of both important functional variables, the U_{wall}-value and AGAW. In that manner, it can be used by architects to obtain a very fast and simple estimation of the expected energy loads when designing a building with a precisely defined thermal transmittance value of the exterior wall elements (U_{wall}-value) and a selected share of the glazing area in the south-oriented façade (AGAW).

4.3.2 Influence of the Building Shape

Designing timber-frame houses with enlarged glazing surfaces offers numerous possibilities of creating structures with a highly attractive shape. Nevertheless, some of the existent studies prove that an attractive and dynamic building design usually results in increased energy demand. The main aim of the current study therefore is to present design solutions where the above increase could be avoided by selecting the appropriate size of the glazing surface in the south façade. Al Anzi et al. [31] investigated the impact of relative compactness (RC) on the annual cooling energy use and the annual total building energy in Kuwait. Their research was based on a prototypical building with over 20 floors and the relevant simulation analysis incorporated several building models with various shapes (L, U, T, cut shape, cross-shape, trapezoid). The impact of the window to wall ratio and that of orientation on the energy use were analysed with respect to various window sizes and glazing types. Furthermore, [32] studied the relationship between the shape and energy requirements during the winter season in two French localities with different climate conditions. They found no correlation between the energy consumption of a building and its shape in a mild climate. Another interesting research is the parametric study by Inanici and Demirbilek [19], whose results show no major influence of the length-to-width ratio on the energy performance of a building when specific climatic conditions are taken into account.

In professional practice, the most used index to describe the shape of the building is the *shape coefficient (F$_s$)* defined as the ratio between the envelope surface of the building (A) and the inner volume of the heated volume of the building (V) given in the form of Eq. (2.2). Albatici and Passerini [33] were encouraged to research new indicators of the energy performance within mild and warm climate conditions related to the building shape. They presented heating requirements of buildings with different shapes in the Italian territory. Their research based on a monthly method (simplified approach) confirms that compactness is more important in cold localities. Hence, the introduction of a new simplified index, the *south exposure coefficient (C$_{fs}$)*:

$$C_{fs} = \frac{S_{south}}{V} \qquad (4.2)$$

where S_{south} represents the surface of the south-oriented façade. Having taken heating requirements into account, Albatici varied envelope areas of the models, while the volume and the percentage of the glazing remained constant.

Another important parameter often used to determine solar access of a building, assuming that the latter is of a given height and optimally oriented, is the aspect ratio (AR), a ratio between the building's length and width (AR = L/W), already presented in Sect. 2.5.1. All the parameters are schematically presented in Fig. 2.8. The aspect ratio is a significant parameter in energy-efficient design concerning the building shape, as emphasized in several studies. In cold climates, for example, the ideal aspect ratio for a rectangle-shaped solar house design ranges from 1.3 to 1.5 [34]. Hachem et al. [35] investigated the effects of the geometric shapes of two-storey single-family housing units on their solar potential by using the so-called *depth ratio a/b*, where *a* represents the length of the shading façade and *b* the length of the shaded façade (Fig. 2.9). The paper demonstrates that both parameters control the extent of shading and consequently a reduction in the solar radiation incident on the shaded facade. It is therefore desirable to reduce the depth ratio in order to optimize the solar potential of façades. A rectangle, with the aspect ratio of 1.3, serves as the reference.

The main aim of the numerical study to be presented is to establish the optimal building shape factor (F_s) with the optimal size of the south-oriented glazing surface for residential timber buildings from the point of view of the energy performance. Two different locations with different macroclimate conditions and solar potential are analysed. The study, solely limited to timber construction, analyses the exterior wall elements in passive design, assuming that the rest of the parameters, such as active technical systems, roof and floor slab assemblies, remain constant. Many of the findings can be beneficial to architects in designing new timber buildings or renovating the existing ones and serve as systematic guidelines for a quick estimation of the building's energy performance, even in the case of highly attractive timber-glass building shapes.

4.3.2.1 Parametric Study: Simulation Model

The presented numerical research is based on a case study of a 3-metre-high one-storey house built in the prefabricated passive timber-frame structural system. The ground floor area $(A_f = 81\ m^2)$, the heated volume $(V = 243\ m^3)$ and the percentage of the glazing size in the south façade (AGAW) are kept constant. The calculations are done for the value of $AGAW_{opt} = 0.35$, which is the optimal size of the glazing placed in the south façade with the external wall standard of $U_{wall} = 0.10\ W/m^2K$. A window glazing (Unitop 0.51–52 UNIGLAS) with three layers of glass, two low-emissive coatings and krypton in the cavities for a normal configuration of 4E-12-4-12-E4 is installed. The glazing configuration with a g-value of 52 % and $U_g = 0.51\ W/m^2K$ assures a high level of heat insulation and light transmission [16]. The window frame U-value is $U_f = 0.73\ W/m^2K$.

On the other hand, the building design and consequently its shape factor $(F_s = A/V)$ vary parametrically. Variations of square, rectangular, L and U shapes of the building were analysed with a total of 8 different ground floor shapes (Fig. 4.21). The building shape factor (F_s) varies parametrically from 1.47 to 1.76.

Climate data for two cities located in different climate conditions, Ljubljana and München, were taken into consideration with a view to getting feedback on the influence of the building shape exposed to different solar radiation. The computer-based analysis Ecotect [36] was used to perform the calculations.

4.3.2.2 Parametric Study: Results and Discussion

The calculated numerical results for the transmission losses (Q_t) through the external walls and the solar gains (Q_s) through the glazing placed in the south façade are presented for both locations, München and Ljubljana (Fig. 4.22).

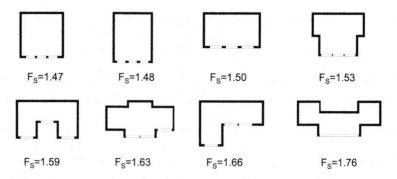

Fig. 4.21 Considered variations of the ground floor shape of the building

The difference between the transmissions heat losses and the solar heat gains increase almost linearly with the shape factor. The reason lies in the fact that the size of the building envelope increases with the dynamics of the building shape. It is, however, important to stress that the function inclination for München is evidently higher than that for Ljubljana.

In order to make further conclusions about the influence of the building shape factor on the annual energy demand of a building, it is important to take into consideration the entire heating and cooling periods. Figures 4.23 and 4.24 therefore show the results for the annual heating and cooling demand for Ljubljana and München. The annual heating demand was calculated by Eq. (2.1) and schematically presented in Fig. 2.2.

Both diagrams point to an interesting dissimilarity between the total annual energy demands for the compared cities, observed at an increased building shape factor. Owing to a lower inclination of the function line presenting the $Q_t - Q_s$ values for Ljubljana (Fig. 4.22) and due to the influence of an increased shape factor value on the cooling demand, the total annual energy demand for the

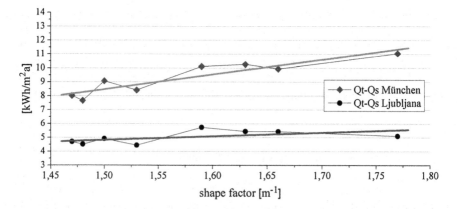

Fig. 4.22 $Q_t - Q_s$ diagrams in dependence on the shape factor for München and Ljubljana

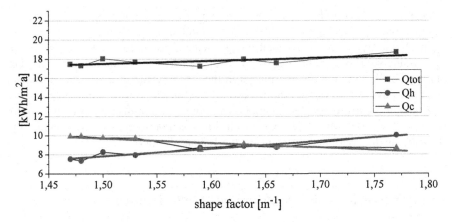

Fig. 4.23 Energy demand for heating (Q_h), cooling (Q_c) and the total energy demand (Q_{tot}) for Ljubljana

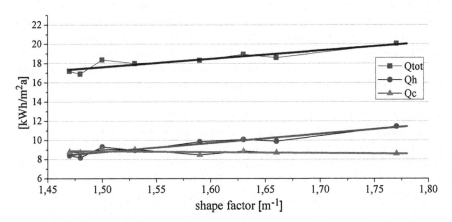

Fig. 4.24 Energy demand for heating (Q_h), cooling (Q_c) and the total energy demand (Q_{tot}) for München

building in Ljubljana is almost independent of the building shape factor. The latter finding is rather contradictory to some well-known existing studies claiming that attractive and dynamic building design usually results in the energy demand increase. The results for the building located in München, on the other hand, are less encouraging (Fig. 4.24) with the influence of the building shape on the total energy demand being far more significant, on account of higher transmission and lower solar radiation in the winter period.

The performed parametric analysis evidently proves that modern passive pre-fabricated timber-frame buildings with the optimal size of the triple glazing placed in the south façade allow for designs incorporating a more complex ground plan and a more attractive design, which results in increasing the value of the building shape factor. This finding is in contradiction to most of the existing studies performed on different buildings shapes. The building in Ljubljana demonstrates an almost constant value of the total energy demand for heating and cooling at an increasing value of the building shape factor, which is due to the fact that the solar heat gains through the glazing placed in the south façade increased in an almost similar way as the transmission losses. Nevertheless, the results show a less encouraging outcome for the locations with lower solar radiation, as compared to our study case of München. Our conclusion resembles some of the findings by Albatici and Passerini [33], given for the Italian territory.

4.4 Structural Stability of Timber-Glass Houses

Modern houses design focuses on ensuring high indoor comfort quality and low-energy consumption. Investors and architects decide upon the orientation of a building and its transparent areas with a view to maximizing the use of natural solar radiation gains. The latter is preconditioned by the appropriate size and orientation of large transparent areas which have to transmit an adequate amount of solar energy into a building in order to provide for natural lighting and heating of the interior space. A comparison of transmission losses through the building envelope with the possible solar heat gains in order to achieve the optimal energy demand of the house is of a great importance in defining the optimal size of glazing areas and a suitable selection of the glazing type.

With respect to the above facts, discussed and analysed in Sect. 4.3, a major part of the glazing needs to be installed in the south-oriented façade for the purpose of better energy performance of a building, which leads to specific technical challenges in the field of structural behaviour of the load-bearing wall elements with an enlarged glazing size. Even though energy-efficient, such construction systems can be extremely problematic when the building is exposed to horizontal loads.

Two of the typical horizontal load cases are wind and earthquake. Their load distribution through the building is the same (Figs. 3.39 and 3.40) although the effects on the building and the subsequent consequences prove to be different. The earthquake, which implies a rapid high intensity dynamic load on the building leading to catastrophic outcomes, is definitely a more concern causing horizontal load case.

One of the basic principles incorporated in designing a building to withstand seismic loads is avoiding *plan irregularity*, i.e., providing for a uniform lateral stiffness distribution in a building. Hence, we eliminate the risk of unfavourable torsion effects which occur with dynamic loads acting upon an irregular floor plan and increase the seismic load on the structure. Energy-efficient buildings with enlarged glazing areas predominantly placed on southern façades may therefore make proper seismic design of the building become a rather difficult and problematic task.

In addition, transparent glass areas usually provide hardly any horizontal stiffness and cause substantial decrease in structural stability of a single-wall assembly. Method A in Eurocode 5 [59] defines that wall panels containing a door or window opening should not be considered as contributing to the racking load-carrying capacity. Method B is less restrictive and declares that the lengths of the panel on each side of the opening formed in the panel should be considered as separate panels. Nevertheless, the use of the most accurate numerical FEM approach (Sect. 3.3.1) reveals that wall panels containing a door or window opening decrease the racking resistance and significantly lower the horizontal stiffness of prefabricated frame-panel wall elements on the one hand, but can still contribute to the horizontal stability of the entire wall assembly on the other. The decreasing factor for the horizontal resistance and stiffness depends on the size and position of the openings.

Section 3.4 discussed a problematic area relating to the horizontal stability of multi-storey timber-frame buildings and presented various options of strengthening prefabricated wall elements in lower storeys where the horizontal load impact is the highest (Fig. 3.52). The horizontal load at the top of the wall element is shifted over the connecting plane and the sheathing board to the support (Fig. 3.46). The sheathing board thus assures the horizontal stability of the entire element.

The main idea of using fixed glazing in prefabricated timber-frame-panel wall elements is to replace the classical sheathing boards with glass panes, as seen in Fig. 4.25.

Contribution of glass areas to the stiffness has been so far rather neglected, which can be accounted for by a relatively low strength and ductility levels of ordinary glass used for windows. However, the introduction of modern glass materials, such as tempered and laminated glass or glass-fibre-reinforced polymers

Fig. 4.25 Timber-glass prefabricated walls—replacing the classical sheathing boards with the glass panes

(GFRP) along with the improvements of glass products' strength properties has allowed for the use of large glazing surfaces since they now contribute to the horizontal stiffness and resistance of the wall elements. The function of sheathing boards is thus taken by glass panes whose stiffness assures the horizontal stability of the wall element. The horizontal point load acting at the top of the element is consequently transferred to the supports in the same manner as already presented in Fig. 3.46:

• The adhesive takes over the shear stresses in the gluing line.
• The tensile diagonal of the glass pane shifts the force to the support.

Application of glass panes acting as load-bearing structural elements in the in-plane stress distribution assures the horizontal stability of the building and replaces the usage of visible diagonal elements (Fig. 4.26). Stability problems which can appear in the case of lower-storey wall elements with large glazing areas are solved by the use of steel diagonals. The latter is a common engineering practice to assure stability of the building against horizontal load actions (Fig. 4.26). Inserting diagonal steel elements is often seen as a less desirable option as it tends to cause heat bridges and turns the erection of the building into a complicated manoeuvre. A timber frame with steel connections is another solution, although not an ideal one since it requires large quantities of steel and provides little stiffness in comparison with full wall segments.

One of the main disadvantages of glass as a load-bearing material is its brittleness. Appropriate seismic design relies on the ductility of materials in order to dissipate the earthquake energy and avoid brittle mechanism failures. Notwithstanding the above, although brittle, modern glass proves to be a very strong material with high compression and tension strengths. If properly connected to the framing system via ductile connectors, it could form a potential lateral stiffness system with the capacity of withstanding earthquakes.

Combining timber and glass to make an appropriate load-bearing element is a very complex process involving a combination of two materials with different

Fig. 4.26 Horizontal load distribution in a multi-storey timber-glass building

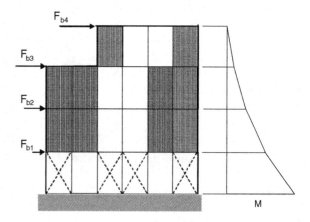

material characteristics, described in Sects. 3.1 and 4.1. The external timber-glass wall elements will be mostly placed in the southern façade of the building and thus exposed to extreme temperature changes, in view of which it is important to stress the different coefficients of thermal linear expansion (α_t) of glass and timber in the grain direction (with the coefficient of glass being two times higher than that of timber). The coefficient of thermal expansion for timber perpendicular to the grain is as much as ten times higher than the coefficient for timber parallel to the grain direction [37] and therefore almost twenty times lower than that of glass. Consequently, an increased temperature effect may increase the shear stress appearing in the adhesives between both materials. Furthermore, design of timber-glass wall elements functioning as load-bearing elements against seismic loads needs to focus on assuring the static resistance of the building in addition to that of its high ductility. Another vital matter to be discussed is an almost complete absence of the remaining load capacity after the appearance of the first crack in the glass panes, which opens a question of finding adequate balance between strength and flexibility of such composed elements. Hence, there is a need for the development of a ductile connection system between timber and glass, a system which can assure high static resistance and take over static loads only (dead load, live load, snow, etc.). According to the presented facts, we can conclude that there is a certain "game played" by the strength and ductility, while the boundary conditions between timber and glass can be classified as having the most important role in the design of such composed elements.

There are also other parameters which can exert significant influence on the horizontal resistance and stiffness of the timber-glass wall elements, such as material characteristics and the thickness of glass panes, even though the boundary conditions between timber and glass tend to be classified as the most important. The above boundary conditions in wall elements are subject to the following most influential parameters:

- Position of the glass pane and consequently the position of the glue line
- Type of the adhesive
- Thickness and width of the glue line.

The three parameters have already been studied by several authors, experimentally and numerically, and will therefore be only briefly presented.

4.4.1 Experimental Studies on Wall Elements

Glass panes were more often used in steel frame systems than in timber-frame systems to assure horizontal stability of the buildings, which calls for a brief overview of some important findings from studies performed on steel frame elements. The emphasis of our discussion will be laid on the conclusions which can be adopted also for the timber-frame wall elements. In the buckling test on the square panes with a four-sided support exposed to a continuous linear compression in-plane load, [38] proves that the ultimate flexure tensile strength of glass depends on the duration of the exposure to load, on surface damage of both panes and on the level of the residual stress during the tempering process. Since the panes are made of heat-strengthened laminated glass, the main parameter for the buckling capacity of the laminated glass is shear stiffness of the PVB interlayer foil, with the remaining two parameters being the thickness of glass and initial surface imperfections.

A possibility of replacing compression elements with glass panes acting as elements of stabilization in modern shell structures was presented by Wellershoff [39]. Using analytical models and an experiment, two different structural systems were developed. System A represented a hinged metal frame with a lever and discrete joints in the frame nodes while system B represented a glass pane which was adhesive bonded to both sides of the metal frame and functions as the shear wall. The adhesives applied were acrylates and polyurethanes, with glass being laminated and heat-strengthened. The two systems were subjected only to in-plane loads or to the combination of the in-plane and out-of-plane loads. System B activated tensile diagonals, in addition to revealing three areas of the highest tension, i.e., in the middle of the glass pane along the tensile diagonal, at the corners of the glass pane—at the starting and the end points of the compression diagonal and finally, at the anchoring point of the adhesive bonded joint. His experimental researches additionally focused on the influence exerted by the duration of exposure to load and by environmental situations, such as UV radia-tion, humidity and temperature. The adhesives applied were silicones, acrylates, polyurethanes and epoxy. Wellershoff came to a conclusion that the shear stiffness of polyurethanes and acrylates was higher than that of silicone. On the other hand, with the increase in temperature, the shear stiffness of polyurethanes and acrylates was decreased while the stiffness of the silicone remained unaltered. Weller [40] tested the same adhesives as Wellershoff. Adhesive bonding of glass for con-struction purposes is feasible in practice under the condition of acquiring con-sensus permits for structures and joints of this kind, ascertains the author.

Adhesive bonding of insulating glass to be installed in winter gardens, façades, residential buildings, etc., was studied by Schober et al. [41]. The test specimens measuring 1.25×2.5 m represented a double insulating glass plate linearly adhesive bonded to the timber frame using acrylates and silicone adhesives. An interesting example of testing glass panes in the steel frame subjected to in-plane and out-of-plane loads can be seen in the research by Močibob [4]. Two concepts of lateral and vertical in-plane load shifts along with a continuous linear out-of-plane load shift were studied. The first was a point support concept with the second being a linear support concept. Both concepts successfully underwent testing as well as exposure to in-plane and out-of-plane loads. The glass used was heat-strengthened and laminated, bonded with construction silicone. According to the author's findings, the lateral in-plane stiffness increases proportionally to the higher thickness of glass and the pane fails in the compression diagonal since the tensile diagonal can no longer support the compression diagonal. Furthermore, Močibob asserts that in-plane and out-of-plane displacements prior to failure are rather high. Peripheral bonding of glass panes onto a metal frame was studied also by Huveners [42] who produced experimental, analytical and numerical proof of the possibility of using such bracing elements in glass façades and single-storey buildings. The test specimens were made of square-shaped toughened glass with a thickness of 12 mm and the size of 1.0×1.0 m. Having developed three different models according to the type of adhesive, the author finds out that epoxy adhesives prove to be more suitable than polyurethanes as their use helps to attain higher in-plane stiffness.

One of the first instances of using glass panes as load-bearing elements in combination with *lightweight timber structures* was presented by Niedermaier [43], according to whom glued joints can be normally classified into three different types (Fig. 4.27). Joint type 1 is a polyurethane or silicone end joint, joint type 2 is a two-sided epoxy joint and joint type 3 is a one-sided epoxy joint. Generally, joint type 2 demonstrates larger stiffness than joint types 3 and 1.

Niedermaier experimentally studied the shear strength of glass panel elements in combination with timber-frame constructions. He tested stiffening glass panel elements which were 800 mm wide and 1,600 mm high. The glass pane was fixed to the timber frame using a joint type 3 with the glue line dimensions of 12-mm-wide and 6-mm-thick polyurethane or silicone adhesive. A horizontal load of 1 kN was applied on the top member. The research results show that the deformability of the timber frame and the tension distribution in the glass depend on the geometry of the adhesive bonded joints as well as on the type of adhesive.

A number of studies on combining glass with timber and those on the in-plane load-bearing capacity of glass panes in *timber-frame wall elements* have been so far carried out by Holzforschung Austria [44] and the Technical University of Vienna [45, 46], which will be of assistance in the comparison with our experimental results. The glazing placed on the external side of the timber frame was not directly glued to the timber frame but bonded with adhesives to the special sub-structure (Fig. 4.28) which is fixed with bolts to the external side of the timber frame. The entire system was protected by the patent HFA Pat.-Nr. 502470. The most important technological advantage of such type of connection is a relatively simple replacement of the glazing replacement in the case of its breakage.

Joint 1   Joint 2   Joint 3

Fig. 4.27 Adhesive joint types presented by Niedermaier [43]

Fig. 4.28 Connection of the glazing to the substructure and the timber frame in HGV elements, adopted from Neubauer and Schober [44], Holzforschung Austria, Pat.-Nr. 502470

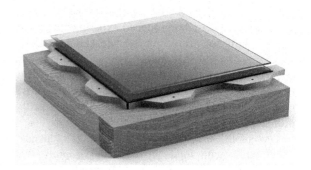

In the experimental analysis, Silicone A with the shear modulus $G = 0.37$ MPa and the glue line dimensions of 14 and 19 mm/3 mm in addition to acrylate with the shear modulus $G = 2.0$ MPa and the glue line dimensions of 14 mm/2 mm were used. A single float-glass pane with a thickness of 8 mm and outer dimensions of $1{,}250 \times 2{,}500$ mm was used to assure the horizontal resistance of the tested elements. Timber-frame elements with dimensions of the cross section of 60/160 mm were composed of timber class GL 24 h. The elements were tested with the horizontal point load acting at the top of the wall and supported by two supports, the tensile and the compressive (Fig. 4.29). The cyclic load procedure (0.1–0.4–1.0 F) according to [47] was performed.

The results for all types of adhesives are presented in Table 4.8. The results measured in the test samples with classical OSB 3 sheathing boards are given for comparison purposes only.

Fig. 4.29 The HGV test samples were subjected to the horizontal point load at the top edge [45]

Table 4.8 Experimental results [44, 45]

Type of the adhesive	Dimensions of the glass panes [mm/mm]	Number of the test samples	Dimensions of the glue line [mm/mm]	Failure force F_u [kN]	Force at $u = h/500$ F_h [kN]	Horizontal stiffness K_h [N/mm]	Slip in the glue line [mm/kN]
Silicone A	1 × 1,250/2,500	$n = 5$	14/3	13.39	2.98	594	0.22
Silicone A	2 × 1,250/2,500	$n = 2$	14/3	24.74	6.76	1,347	0.09
Silicone A	1 × 1,250/2,500	$n = 3$	19/3	21.95	5.12	1,022	0.12
Acrylate	1 × 1,250/2,500	$n = 6$	14/2	38.68	0.62	2,885	0.006
OSB 3	1 × 1,250/2,500		/	34.50	1.30	1,358	/

Although the Silicone A test samples with the 14-mm-wide glue line demonstrated the average failure force at $F_u = 13.39$ kN, it is important to stress that the force occurring at the horizontal displacement of $u = h/500 = 5$ mm was only $F = 2.98$ kN. We can therefore conclude that the horizontal stiffness of the test samples was extremely low. The failure force of the test samples rapidly increased with the width of the glue line. The test samples with a 19-mm-wide glue line demonstrated the average failure force at $F_u = 21.95$ kN, which meant an increase of 64 %. The value of the force at serviceability limit state condition demonstrated a more rapid increase towards the value of $F_h = 5.12$ kN and a subsequent increase in stiffness by 72 %.

It is furthermore interesting to compare the results of the test samples with a single glass pane with those having two glass panes, both placed on the external sides of the timber-frame elements. The results demonstrate an increase in the failure force of 85 % and an increase in the stiffness of 127 %. This finding can be of assistance in designing timber-frame-panel multi-storey buildings located in heavy windy or seismic areas.

As described at the beginning of the chapter, a general construction-related goal is to replace the classic sheathing boards (wood-based or fibre-plaster boards) with glass panes. A comparison of the measured results obtained for the wall elements with glass panes with those relevant to the elements with the classical OSB boards witnessed a considerable reduction in the load-bearing capacity and stiffness. The failure force of the test samples where the Silicon A adhesive with a 14-mm-wide glue line was applied exhibited merely 39 % of the failure force of the element with the OSB boards. The stiffness underwent a similar, 44 % reduction.

Appropriate resistance of the wall elements with glass panes demands application of acrylate adhesives which make the slip in the glue line evidently smaller than silicone adhesives. The load-bearing capacity and the stiffness in particular were noticeably higher than in elements with OSB boards. Nevertheless, using acrylate adhesives may lead to problems related to the ductility of the connection and the consequent seismic resistance of the bearing elements, in addition to potential relative deformation of both connected materials under a strong temperature effect.

A fact that needs to be underlined is the existence of first timber buildings with HGV timber-glass elements (Austria) functioning as resisting wall elements under horizontal load and resisting the horizontal stiffening of the building. The most interesting is probably a low-energy two-storey single-family house in Eichgraben (Austria), Fig. 4.30. The timber frames of the wall elements were factory-made and transported to the building site where the glass elements were fixed to the timber frame using the HFA Pat.-Nr. 502470 type of connection [44].

The second building realized within the Holzforschung Austria (HFA) research project was a bungalow with a north-facing glass façade and a south-facing glass façade. The building revealed a feasible assembly and manufacture of the cladding system and offered open space for the future insights in durability and long-term behaviour of buildings [11]. As opposed to the family house in Eichgraben, the

Fig. 4.30 Two-storey single-family house with HGV wall elements, built in Eichgraben (Austria)

wall elements produced in a multi-panel timber-frame system were finalized in the factory and transported to the building site.

In the study by Blyberg [10], a shear wall element intended to be used as a load-bearing façade element was designed. The laminated-float-glass pane with a thickness of 10 mm was placed in the middle of the timber frame (Fig. 4.31). LVL (laminated veneer lumber) with a machined groove squared cross sections of 45 mm was used for timber elements. Timber was glued onto glass with the acrylate adhesive Sikafast and in a few cases, with the 2-component silicone-based adhesive Sikasil SG-500. The edge of the glass was thus visible. Three different load cases were used for both adhesive types; horizontal load, vertical load and a combination of horizontal and vertical loads. The elements were subjected to the horizontal point load and supported with two supports, the tensile and the compressive, in a similar way as in Neubauer and Schober [44] experiments.

The obtained maximal horizontal load for all tested elements was 41.4 kN for the silicone specimens and 67.8 kN for the acrylate specimens. While the results from the adhesive testing showed that the acrylate adhesive had much larger strength than the silicone adhesive, it should be noted that the acrylate has a glass-transition temperature of 52 °C, which could imply that the properties of the adhesive change at increased temperatures. It is also interesting to compare the results with the values obtained by Neubauer and Schober [44] tests. The reason for significantly higher lies in the fact that the bond line was only 1.5 mm thick, which is two times lower than in the case of the silicone type in Neubauer and Schober [44].

Fig. 4.31 Test configuration
of the wall test specimens

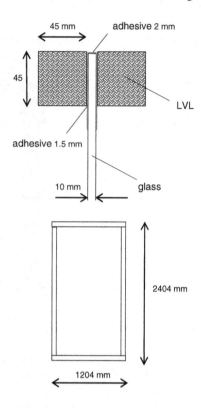

At the University of Minho, a new timber-glass panel element was developed which can be applied either as a slab (Fig. 4.32a) or a wall (Fig. 4.32b) prefabricated load-bearing element. According to its dimensional metrics, it is adjustable

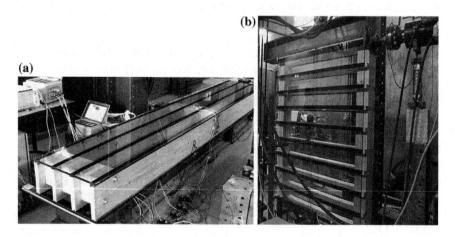

Fig. 4.32 Timber-glass panel **a** as a slab element [6], **b** as a wall element [48]

to several foreseen project situations. In the huge experimental analysis [6], twenty one panels were tested—eleven timber panels and ten timber-glass composite panels. Each composite panel was 224 mm thick and consisted of two laminated glass panes bonded on both faces of the timber structure, made of four Pinus Sylvestris timber boards, with a cross section of 200 mm x 30 mm. The specimens were tested in bending as slab elements and as wall elements subjected to vertical load.

The main conclusion to withdraw from this experimental work was that glass behaves as structural reinforcement of the timber substructure, particularly when used as a *structural slab element*, in which case the tests results showed excellent structural performance of the composite panel with an increase of 31 % in the maximum load obtained, in comparison with the glass-less panel.

As a *structural wall system* tested under vertical load, the contribution of glass became even more evident. The bearing capacity of the timber-glass composite panels was compared to that of timber panels without glass. The results showed a clear increase in the stiffness and resistance, which allowed the value of 100 kN to be exceeded, while still keeping a considerable safety margin and ductile failure at its post-high peak.

The following step was to develop several implantation models either as semi-detached houses or blocks in order to produce an innovative timber-glass composite construction system in which the combination of timber and glass simultaneously incorporates energetic, functional and aesthetic characteristics. Such system becomes an architectural and structural skin, a frontier between the inner and outer spaces reinforcing the importance of the structure's energetic performance and the comfort of its inhabitable space, predominantly in terms of thermal transfers, air circulation and natural lighting levels—features that definitely contribute to optimizing the energy efficiency and effectiveness of its management. The second phase involving optimization of the structural solution, based on the search for tectonics and a contemporary architectural system construction, led to the materialization of the housing model with the above-described load-bearing composite timber-glass slab and wall elements (Fig. 4.33).

A set of experimental tests on timber-frame-panel wall elements were also performed at the University of Maribor, in 2011 and 2012. The tests were subdivided into two main groups according to the position of the glass panes:

• Glass panes were placed on the external sides of the timber frame, Fig. 4.35 [49].
• A single glass pane was embedded into the middle plane of the timber frame, Fig. 4.38 [50].

The test specimens consisted of a timber frame with the outside edges measuring 1,250/2,640 mm (Fig. 4.35), which used to be a standard size of wall panels tested in previous studies where a different sheathing material was used, see Sects. 3.3 and 3.4. Vertical studs were composed of rectangular 90/90-mm timber elements with the size of horizontal girders being 90/80 mm. The bottom left-hand

Fig. 4.33 Application of prefabricated elements in a housing model [6]

corner of the panel had three 16-mm holes through which the panel was fixed into the stirrup functioning as the tensile support. Timber-frame elements in both cases were made of wood with a strength grade C22, glass panes consisted of toughened ESG glass, and the adhesive used in the timber-glass joint was a two-component silicone adhesive type Ködiglaze S, produced by Kömmerling [51]. In the second case, polyurethane and epoxy adhesives were additionally used. Material properties of timber with a strength grade C22 were taken from [52], with properties of thermally toughened glass being taken from [53] and [54] and those of adhesives obtained from the producer's technical sheet [55, 56]. All material properties are listed in Table 4.9.

Testing procedure

After a relaxation period of several days, the panels were installed in the static load testing machine. During the testing process, the panels were rotated by 90° and fixed with the left vertical stud via three coil bars Φ16 into the stirrup consisting of two steel plates, as shown in Fig. 4.34a. The reaction of the lower compression support was taken by the steel section I180, fixed to the stiff steel frame. The test specimens were exposed to force F which approximates lateral load, ranging from point 0 to the failure point according to the [57] static monotonic testing procedure, Fig. 4.34b. The load stress increase rate on the

Table 4.9 Properties of the materials used

	$E_{0,m}$ [N/mm²]	G_m [N/mm²]	$f_{t,0,k}$ [N/mm²]	$f_{m,k}$ [N/mm²]	$f_{v,k}$ [N/mm²]
Timber C22	10,000.00	630	13.0	22.0	2.4
Thermally toughened glass EN 12150	70,000.00	28,000.00	45	120	/
2C Silicone adhesive (Ködiglaze S)	2.8	0.93	2.1	/	3.15
Polyurethane adhesive (Ködiglaze P)	1.0	0.33	2.0	/	2.0
Epoxy adhesive (Körapox 558)	26.0	8.58	28.59	/	22.0

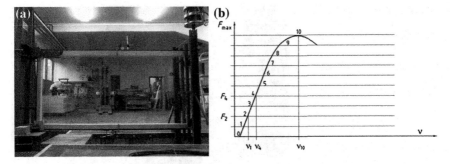

Fig. 4.34 Test configuration (**a**), and the load-testing procedure according to [57] (**b**)

panels was 2.0 kN/250 s for the values from 0 to 10 kN and 2.0 kN/200 s for the value of 10 kN to the point of failure.

4.4.1.1 Glass Panes Placed on the External Sides of the Timber Frame

The first step of the study involved glass panes with a thickness of 6 mm placed on the external sides of the timber frame. The type of connection resembled Joint 3 presented by Niedermaier [43], Fig. 4.27, but featured an important difference

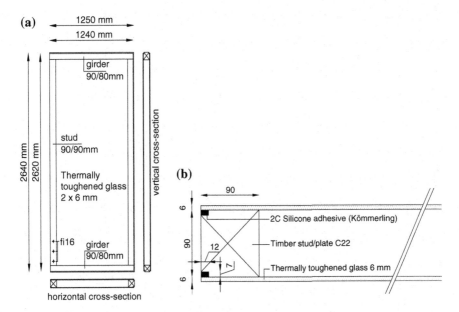

Fig. 4.35 Dimensions of the test specimens (**a**), and the section of the linear adhesive bonded joint (**b**)

Fig. 4.36 Brittle failure
mode of the glass panes in
compression

seen in the silicone adhesive being applied into a special groove in the timber
frame (Fig. 4.35). The adhesive used had a thickness of 7 mm with the width of
the glue line measuring 12 mm. An important part of the research focused on the
impact of the relaxation time of the silicon adhesive, from the time of bonding to
the starting point of the test specimen exposure to load. Consequently, the test
specimens with the longest relaxation time ($t = 7$ days) were given the label
ST-01, those with the shortest relaxation time ($t = 3$ days) were labelled as ST-02,
and finally, the specimens with the middle relaxation time ($t = 5$ days) received
the ST-03 label.

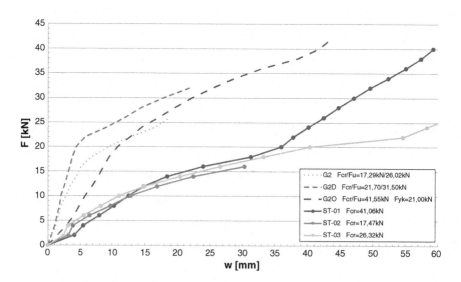

Fig. 4.37 F-w diagrams of the test specimens with different types of sheathing material [49]

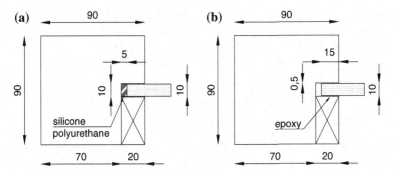

Fig. 4.38 Connection with a glue line in the middle of the timber frame **a** silicone and polyurethane adhesive, **b** epoxy adhesive

The behaviour of the tested samples was very similar and it proved a completely brittle failure mode of both external glass panes, occurring under the compressive stress in the glass pane, Fig. 4.36.

The results of all specimen groups are given in Fig. 4.37 which shows the normalized values of vertical displacements (w) relative to force F, separately for each specimen. The F_{cr} values given in the legend to the figure demonstrate the force at which the first crack appeared in the glass sheathing. Owing to the non-ductile behaviour of glass (Fig. 4.38), the latter force also meant the failure force. Figure 4.39 presents diagrams of the normalized mean values of displacements of

Fig. 4.39 Failure modes of the samples with a glass pane placed in the middle of the timber frame; **a** destruction of the timber corner in the case of using silicone and polyurethane adhesives, **b** a brittle glass rupture in the a case of epoxy adhesive

the test specimens, taken from our experimental study presented in Sects. 3.3 and 3.4, where the test specimens with identical geometrical characteristics as those in the present research had different sheathing materials. The glass panes are labelled as ST while the labels of other test specimens mean the following:

- G2—single FPB sheathing with a span of $s = 75$ mm between the staples
- G2D—double FPB sheathing with a span of $s = 75$ mm between the staples
- G2O—single OSB sheathing with a span of $s = 75$ mm between the staples.

The stiffness in the F-w diagram is defined by the inclination of the curve. The load-bearing capacity is defined by the value of the failure force.

The test specimen's behaviour demonstrated its dependence on the age of the silicone adhesive at the time of exposure to load. Subsequently, the behaviour of the test specimen ST-01 with the longest relaxation time ($t = 7$ days) proved to be the best and could be labelled as three-linear (phase 1: linear behaviour until the point of yielding of the adhesive; phase 2: yielding of the adhesive; phase 3: fixing in the connecting plane—the glass sheathing leans on the timber frame, which is followed by instantaneous failure of the glass sheathing). On the other hand, the test specimen ST-02 with the shortest relaxation time ($t = 3$ days) failed soon after the onset of the yielding of the adhesive.

Although the test specimen ST-01 can be compared to G20 in its load-bearing capacity, its stiffness is nevertheless essentially lower. A similar comparison of the load-bearing capacity can be made between the test specimens ST-03 and G2, where the latter displays approximately three times higher stiffness in the linear-elastic behaviour range. The above conclusions bear a close similarity to the findings presented by Niedermaier [43] Neubauer and Schober [44] and Cruz and Pequeno [6].

4.4.1.2 Single Glass Pane Embedded into the Middle Plane of the Timber Frame

Another possibility of using glass panes as a load-bearing sheathing material in timber-frame wall elements is to insert a single glass pane into the middle plane of the timber frame. A similar type of connection was already presented in the study by Niedermaier [43] (Fig. 4.27), Joint 1 and Joint 2. Our study comprised testing of three groups with different boundary conditions and adhesives:

- Silicone adhesive with a glue line thickness of 6 mm placed in the lateral plane of the connection (Fig. 4.38a).
- Polyurethane adhesive with a glue line thickness of 6 mm placed in the lateral plane of the connection (Fig. 4.38a).
- Epoxy adhesive with a glue line thickness of 0.5 mm placed in the shear plane of the connection (Fig. 4.38b).

Each test group consisted of three specimens. The aim of the study was to compare the results relative to application of different adhesives types, with a special focus on the obtained load-bearing capacity and stiffness after using an elastic adhesive (e.g., silicone) or a stiff adhesive like epoxy. Another point of analysis was the influence of the adhesive type on the ductility of the test samples. A single fully tempered (toughened) glass pane with a thickness of 10 mm having the material characteristics given in Table 4.9 was used for all test samples.

It is furthermore interesting to compare the behaviour of the tested samples with different boundary conditions and different types of the adhesives. The failure modes of the elements with elastic adhesives (silicone and polyurethane) demonstrated a relatively ductile failure with destruction of the corner connection between the timber elements (Fig. 4.39a). On the other hand, failure of the test samples with a very stiff adhesive (epoxy) was completely brittle with a glass rupture in the compressive diagonal (Fig. 4.39b). Both findings prove the Cruz et al. [7] conclusions.

Figure 4.40 presents diagrams of the normalized mean values of displacements of all tested samples with glass panes in addition to the results of the test specimens with identical geometrical characteristics but with different sheathing materials. The results obtained on silicone HGV test samples with the 14/3 mm (HGV 14/3) and 19/3 mm (HGV 19/3) glue lines, taken from Neubauer and Schober [44] and Hochhauser [45], merely serve to provide further comparison.

Similar results of the test samples with silicone and polyurethane adhesives prove to be an important piece of information relevant to a better overall technological advance of silicone adhesives, presented in Sect. 4.1. Silicone samples'

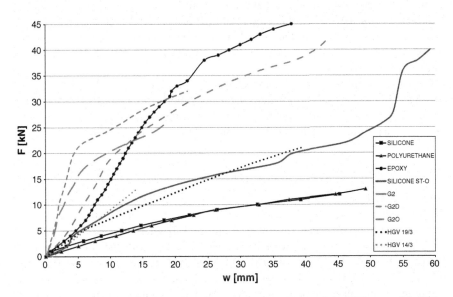

Fig. 4.40 F-w diagrams of the test specimens with different types of sheathing materials [50]

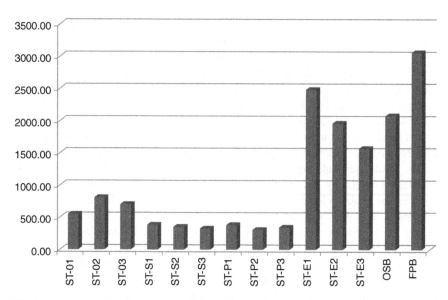

Fig. 4.41 Calculated results for the racking stiffness according to [57]

results are comparable to those by Neubauer and Schober [44] with a difference in the glue line dimensions being taken into account. Moreover, the failure force of the silicone and polyurethane test samples is set at about of 40–50 % of the values of typical classical sheathing materials (fibre-plaster boards or OSB). Neverthe-less, a comparison with the results for the force forming the first crack (Table 3.12) in the fibre-plaster boards (F_{cr}) shows that the silicone adhesive specimens reached almost 70 % of the value obtained on the FBP test samples with a typical 75-mm span staple disposition. On the other hand, the results for the failure force of the epoxy adhesive test samples prove to be in an absolutely comparable range with those of the OSB boards.

As mentioned beforehand, the connection between glass and timber has the strongest influence on the stiffness of the wall elements, which is also evident from the results in Fig. 4.41 and from those for the racking stiffness (R) of all tested wall elements with glazing, calculated according to the [57] in the prescribed form of

$$R = \frac{F_4 - F_2}{w_4 - w_2} \tag{4.3}$$

and graphically presented in Fig. 4.41. The values for the elements with OSB and FPB boards are given for the purpose of the comparison of the stiffness.

The results of the mean values are found in Table 4.10.

The racking stiffness (R) for all glass test samples, with the exception of the epoxy type, was clearly far under the stiffness of the classical load-bearing wall elements with OSB or FPB sheathing boards. We can therefore point out once more that using glass elements instead of classical sheathing boards exerts a more

Table 4.10 Measured experimental results (mean values)

Type of the test samples	R [N/mm]	F_{cr} [kN]	F_y [kN]	$F_{u,k}$ [kN]
OSB 3 ($s = 75$ mm)	2,078.12	/	21.25	41.55
FPB ($s = 75$ mm)	3,059.39	17.06	/	26.17
ST-0 (silicone ext. side)	686.76	/	16.80	28.28
ST-S (silicone middle)	354.48	/	11.58	11.58
ST-P (polyureth. middle)	344.82	/	13.43	13.43
ST-E (epoxy middle)	2,004.41	37.35	/	37.35

F_{cr}— force forming the first crack in the sheathing board or glass pane
F_y—force forming the fasteners or adhesive yielding (or timber-frame corner destruction)
$F_{u,k}$— ultimate failure force (corner destruction or glass failure)

rigorous influence on the racking stiffness than on the load-bearing capacity of the wall elements, which consequently leads to more problems in satisfying the serviceability limit state requirements. The above findings are closely related to the conclusions drawn in Sect. 3.3.1, discussing the problem of non-resisting window and door openings in wall elements.

4.4.2 Computational Models

As presented in the experimental studies, there are many parameters which significantly influence the behaviour of timber-frame wall elements with glass panes exposed to horizontal load. The following are the most important:

- Type of glazing
- Thickness of the glazing
- Type of adhesive
- Dimensions of the glue line.

Designing timber-glass wall elements is thus a very complex process with the most important fact being the approximation of the timber-glass connection in the bond line. Since there is usually a shortage of time and funds to perform experimental analysis of the elements to be used, it is of utmost importance to develop appropriate computational models which will serve as means of accurate prediction of the horizontal load-bearing capacity and the horizontal stiffness of such wall elements. The models, based on general guidelines of designing timber-frame wall elements and already discussed in Sect. 3.3, can be classified in two main groups:

- Semi-analytical spring models
- Finite element models (FEM).

4.4.2.1 Semi-Analytical Spring Models

The models are based on the main assumptions relevant to the semi-analytical simplified composite beam model in Sect. 3.3.4. A major difference between the two types of models is replacing the timber-frame-sheathing board connection plane with the timber-glass bond line. It is consequently not possible to adopt the so-called γ-procedure used in the above-mentioned computational model to simulate the timber–board connection. The springs are used instead (Fig. 4.42).

The first of such static undetermined spring models was introduced by Kreuzinger and Niedermaier [58]. The timber-frame bending stiffness is supposed to be perfectly rigid (EI $= \infty$). Another approximation is identical stiffness of the springs longitudinal (k_u) and perpendicular (k_w) to the glue line ($k = k_u = k_w$). According to these simplifications and considering the classical beam theory, the maximal horizontal displacement (u) under the horizontal force (F_H) acting at the top of the wall element is developed in the form of:

$$u = \frac{2 \cdot F_H}{k \cdot b} \cdot \left(\frac{1}{1 + \frac{h}{3b}} + \frac{\frac{h}{b}}{1 + \frac{h}{3b}} \right) \qquad (4.4)$$

In the above equation, the stiffness of the bond line (k) is determined in dependence on the shear modulus of the adhesive (G_{bl}), the thickness (d_{bl}) and the width (b_{bl}) of the bond line:

$$k = G_{bl} \cdot \frac{b_{bl}}{d_{bl}} \qquad (4.5)$$

If we consider the shear stress (τ) as uniformly distributed along the bond line, then it can be written in the form of

Fig. 4.42 Spring model introduced by Kreuzinger and Niedermaier [58]

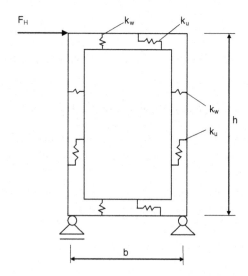

$$\tau = \frac{F_H}{b_{bl} \cdot d_{bl}} \cdot \frac{1}{1 + \frac{h}{3b}} \qquad (4.6)$$

and the slip (Δ) in the bond line between the glass pane and timber-frame elements can be finally developed in the form of

$$\Delta = \frac{\tau}{G_{bl}} \cdot d_{bl} \qquad (4.7)$$

4.4.2.2 Finite Element Models

The process of modelling timber-frame wall elements with glass panes under the horizontal load by using the Finite Element Method is the most accurate but at the same time the most complex and time-consuming approach. In view of the latter, simplified "hand-calculating" methods described beforehand tend to be applied in practice since they provide the user with results without recourse to complex and expensive software.

FEM modelling of the timber-glass composite wall elements is based on the process of modelling timber-frame wall elements with classical sheathing boards (FPB or OSB) with and without openings (c.f. Sect. 3.3.1), schematically presented in Fig. 3.4. *Timber-frame material* is considered as an isotropic elastic material, and the elements of the timber frame are modelled as simple plane-stress elements. The *glass panes* are modelled by using the non-linear 2D shell elements. Since the material behaviour of glass proves to be extremely non-ductile with brittle failure modes, glass is presumed to be linear-elastic until failure. The *bond line* between timber and glass is modelled with 3D layer elements having the material characteristics of the adhesive.

References

1. Rasmussen SC (2012) How glass changed the world, the history and chemistry of glass from antiquity to the 13th century. Springer
2. Staib G (1999) From the origins to classical modernism. In: glass construction manual. Birkhäuser—Publishers for Architecture
3. Wurm J (2007) Glass structures—design and construction of self-supporting skins. Birkhäuser Verlag AG, Basel, Boston, Berlin
4. Močibob D (2008) Glass panel under shear loading—use of glass envelopes in building stabilization. PhD thesis, EPFL, Thèse no 4185, Lausanne, Switzerland
5. Balkow D (1999) Glass as a building material. In: glass construction manual. Birkhäuser—publishers for architecture
6. Cruz P, Pequeno J (2008) Timber-glass composite structural panels: experimental studies and architectural applications. Conference on architectural and structural applications of glass, Delft University of Technology, Faculty of Architecture, Delf, Netherlands

7. Cruz P, Pacheco J, Pequeno J (2007) Experimental studies on structural timber glass adhesive bonding. COST E34, bonding of timber, 4th workshop ≫ practical solutions for furniture and structural bonding, Golden Bay Beach Hotel, Larnaca, Cyprus, 22–23 Mar 2007

8. Haldimann M, Luible A, Overend M (2008) Structural use of glass. IABSE 2008

9. Blyberg L, Serrano E, Enquist B, Sterley M (2012) Adhesive joints for structural timber/glass applications: experimental testing and evaluation methods. Int J Adhes Adhes 35:76–87

10. Blyberg L (2011) Timber/glass adhesive bonds for structural applications. Licentiate thesis by Louise Blyberg, Linnaeus University, School of Engineering

11. Winter W, Hochhauser W, Kreher K (2010) Load bearing and stiffening timber-glass-composites (TGC). WCTE 2010 conference proceedings

12. Pequeno J, Cruz P (2009) Structural timber-glass linear system: characterization and architectural potentialities. Glass performance days 2009, Tampere, Finland

13. UNIGLAS® (2010) College technical compendium, 1st ed., UNIGLAS® GmbH & Co. KG, Montabaur

14. ISO 10077-1:2006 (2006) Thermal performance of windows, doors and shutters. Calculation of thermal transmittance

15. Dow corning insulating glass manual', literature number 62-1374D-01

16. Gustavsen A, Jelle BP, Arasteh D, Kohler K (2007) State-of-the-art highly insulating window frames. Research and market review, Oslo

17. Johnson R, Sullivan R, Selkowitz S, Nozaki S, Conner C, Arasteh D (1984) Glazing energy performance and design optimization with daylighting. Energy Build 6:305–317

18. Steadman P, Brown F (1987) Estimating the exposed surface area of the domestic stock. Energy and urban built form, Centre for Architectural and Urban Studies, University of Cambridge, pp 113–131

19. Inanici NM, Demirbilek FN (2000) Thermal performance optimization of building aspect ratio and south window size in five cities having different climatic characteristic of Turkey. Build Environ 35(1):41–52

20. Bülow-Hübe H (2001) The effect of glazing type and size on annual heating and cooling demand for Swedish offices. (Report No TABK–01/1022). Department of construction and architecture, Lund University, Division of Energy and Building Design, Lund

21. Persson ML, Roos A, Wall M (2006) Influence of window size on the energy balance of low energy houses. Energy Build 38:181–188

22. Persson ML (2006) Windows of opportunities, the glazed area and its impact on the energy balance of buildings. PhD Thesis, Uppsala Universitet

23. Ford B, Schiano-Phan R, Zhongcheng D (2007) The passivhaus standard in European warm climates, design guidelines for comfortable low energy homes—part 2 and 3. Passive-on project report. School of the built environment, University of Nottingham

24. Bouden C (2007) Influence of glass curtain walls on the building thermal energy consumption under Tunisian climatic conditions: the case of administrative buildings. Renew Energy 32:141–156

25. Hassouneh K, Alshboul A, Al-Salaymeh A (2010) Influence of windows on the energy balance of apartment buildings in Amman. Energy Convers Manage 51:1583–1591

26. Praznik M, Kovič S (2010) With active systems and thermal protection to passive and plus energy residential buildings. Published international conference proceedings, energy efficiency in architecture and civil engineering, University of Maribor, Faculty of civil engineering, Maribor, pp 45–57

27. Žegarac Leskovar V, Premrov M (2011) An approach in architectural design of energy-efficient timber buildings with a focus on the optimal glazing size in the south-oriented Façade. Energy Build 43:3410–3418

28. Žegarac Leskovar V (2011) Development of design approach for the optimal model of an energy-efficient timber house. PhD Thesis, Graz University of Technology

29. ARSO (2010) Climate conditions in Slovenia. http://meteo.arso.gov.si/uploads/probase/www/climate/text/sl/publications/podnebne_razmerev_Sloveniji_71_00.pdf, (20.08.2010)

30. Feist V (2007) PHPP 2007 guide book. Passivhaus Institut Dr. Wolfgang Feist Darmstadt

31. Al Anzi A, Seo D, Krarti M (2009) Impact of building shape on thermal performance of office buildings in Kuwait. Energy Convers Manage 50(3):822–828
32. Depecker P, Menezo C, Virgone J, Lepers S (2001) Design of building shape and energetic consumption. Build Environ 36(5):627–635
33. Albatici R, Passerini F (2011) Bioclimatic design of buildings considering heating requirements in Italian climatic conditions. A simplified approach. Build Environ 46(8):1624–1631
34. Chiras D (2002) The solar house: passive heating and cooling. Chelsea Green Publishing, White River Junction
35. Hachem C, Athienitis A, Fazio P (2011) Parametric investigation of geometric form effects on solar potential of housing units. Sol Energy 85:1864–1877
36. Ecotect analysis (2011) Sustainable building design software—Autodesk
37. Hoadley RB (2000) Understanding wood, A Craftsman's guide to wood technology. The Taunton Press, USA
38. Luible A (2004) Stabilität von Tragelementen aus Glas. PhD thesis, EPFL, Thèseno 3014, Lausanne, Switzerland
39. Wellershoff F (2006) Nutzung der Verglasung zur Aussteifung von Gebäudehüllen. PhD Thesis, Schriftenreihe—Stahlbau RWTH Aachen, Heft 57, Shaker Verlag, Aachen, Germany
40. Weller B (2007) Designing of bonded joints in glass structures. In: Proceedings of the 10th international conference on architectural and automotive glass (GPD), Tampere, Finland, pp 74–76
41. Schober KP, Leitl D, Edl T (2006) Holz-Glas-Verbundkonstruktionen zur Gebäudeaussteifung. Magazin für den Holzbereich, Heft 1, Holzforschung Austria, Vienna
42. Huveners EMP (2009) Circumferentially adhesive bonded glass panes for bracing steel frames in facades. PhD thesis, University of Technology Eindhoven, Netherland
43. Niedermaier P (2003) Shear-strength of glass panel elements in combination with timber frame constructions. In: Proceedings of the 8th international conference on architectural and automotive glass (GPD), Tampere, Finland, pp 262–264
44. Neubauer G, Schober KP (2008) Holz-Glas-Verbundkonstruktionen, Weiterentwicklung und Herstellung von Holz-Glas-Verbundkonstruktionen durch statisch wirksames Verkleben von Holz und Glas zum Praxiseinsatz im Holzhausbau (Impulsprojekt V2 des Kind Holztechnologie), Endbericht, Holzforschung Austria, Vienna, Austria
45. Hochhauser W (2011) A contribution to the calculation and sizing of gued and embedded timber-glass composite panes. PhD Thesis, Vienna University of Technology, Faculty of Civil Engineering, Austria
46. Hochhauser W, Winter W, Kreher K (2011) Holz-Glas-Verbundkonstruktionen—state of the art, Forschungsbericht, Studentische Arbeiten. Technische Universitat Wien, Institut fur Architekturwissenschaften Tragwerksplanung und Ingenieurholzbau
47. European committee for standardization (1996) EN 594:1996: timber structures—test methods—racking strength and stiffness of timber frame wall panels. Brussels
48. Cruz P, Pequeno J, Lebet JP, Mocibob D (2010) Mechanical modelling of in-plane loaded glass panes. Challenging glass 2—conference on architectural and structural applications of glass, TU Delft, May 2010
49. Ber B, Kuhta M, Premrov M (2011) Glazing influence on the horizontal load capacity and stiffness of timber-framed walls. In: Proceedings of the 33rd assembly of structural engineers of Slovenia, Bled, 6–7 Oct 2011, pp 301–308
50. Ber B, Premrov M, Kuhta M (2012) Horizontal load-carrying capacity of timber-framed walls with glass sheathing in prefabricated timber construction. In: Proceedings of the 34th assembly of structural engineers of Slovenia, Bled, 11–12 Oct 2012, pp 211–218
51. Kömmerling (2008) Product information Ködiglaze S—special adhesive for structural and direct glazing
52. European committee for standardization (2003) EN 338:2003 E: structural timber—strength classes. Brussels

53. Kömmerling (2008) Product information Ködiglaze P—special adhesive for bonding insulating glass units into the window sash
54. European committee for standardization (2004) EN 572-1:2004: glass in building—basic soda lime silicate glass products—part 1: definitions and general physical and mechanical properties. Brussels
55. European committee for standardization (2000) EN 12150-1:2000: glass in building—thermally toughened soda lime silicate safety glass—part 1: definition and description. Brussels
56. Kömmerling (2011) Product information Körapox 558—two component reaction adhesive for bonding of metals, for example steel or aluminium to each other
57. European committee for standardization (2011) EN 594:2011: timber structures—test methods—racking strength and stiffness of timber frame wall panels. Brussels
58. Kreuzinger H, Niedermaier P (2005) Glas als Schubfeld. Tagungsband Ingenieurholzbau, Karlsruher Tage
59. European Committee for Standardization CEN/TC 250/SC5 N173 (2005) EN 1995-1-1:2005 Eurocode 5: design of timber structures, part 1-1 general rules and rules for buildings, Brussels
60. CSN EN 1279-1—glass in building—insulating glass units—part 1: generalities, dimensional tolerances and rules for the system description

Printed in the United States
By Bookmasters